THE ETHNOGRAPHY OF RHYTHM

Verbal Arts: Studies in Poetics
Lazar Fleishman and Haun Saussy, series editors

THE ETHNOGRAPHY OF RHYTHM

Orality and Its Technologies

HAUN SAUSSY

Fordham University Press

NEW YORK   2016

Visit us online at www.fordhampress.com.

Library of Congress Cataloging-in-Publication Data

Names: Saussy, Haun, 1960– author.
Title: The ethnography of rhythm : orality and its technologies / Haun Saussy.
Description: First edition. | New York : Fordham University Press, 2016. |
    Series: Verbal arts | Includes bibliographical references and index.
Identifiers: LCCN 2015036361 | ISBN 9780823270460 (hardback) |
ISBN 9780823270477 (paper)
Subjects: LCSH: Oral tradition. | Poetics. | Orality in literature. |
    Storytelling. | Folk literature—History and criticism. | BISAC: LITERARY
    CRITICISM / Semiotics & Theory. | SOCIAL SCIENCE / Anthropology /
    Cultural. | TECHNOLOGY & ENGINEERING / Social Aspects.
Classification: LCC GR72 .S28 2016 | DDC 808.5/43—dc23
LC record available at http://lccn.loc.gov/2015036361

Printed in the United States of America

18 17 16   5 4 3 2 1

First edition

À la mémoire de

Anne des Prez de la Morlais
François Desgrées du Loû

Contents

Foreword

Olga V. Solovieva

What do the learning of the Druids, the abbé Rousselot's "speech in-scriber," and Marcel Jousse's little dancing girls have in common? The answer resides in the pocket of any user of a cell phone today. Every "text" we send or receive participates in embodied orality. To be sure, a "text" is made of letters, but letters supplementing what is conventionally known as writing with abbreviations, misspellings, diacriticals, capitals, emoji—introducing hybridity into the alphabet and making it a distance-projection of the gesticulating body.

If this is "secondary orality," it is nonetheless not that predicted by Marshall McLuhan from his observations of the rise of radio and television in the 1950s.[1] Those "post-Gutenberg" media simply recorded and transmitted speech as speech, perhaps increasing the presence of spoken words in our lives but not changing substantially the ontological status of oral versus written communication. Our habits of electronic mediation now tacitly reverse the very episteme that understood orality simply as the absence of writing. Text messaging pulls writing into the orbit of orality while capturing the movements of orality in a "writing machine."

But this new paradigm, as always, is not entirely new. Though orally transmitted, the Druids' sacred knowledge, we learn, was wired into the priestly minds like writing through decades of memorization. Rousselot's phonautograph wrote down individual modulations of speech to capture forms of orality usually treated as peripheral to the system of language. Jousse, the inventor of the so-called rhythmocatechism, interpreted the

Scriptures as a written code accompanying gestural, dancelike recitals, memorized to assure the everlasting immediacy of God's word.

Media archaeology, as exhibited in this book, feeds back into media ecology. All media relate to one another genealogically, as predecessors and successors, and functionally, as alternatives. Étienne-Jules Marey's early recordings of movement, which guided Rousselot's search for a form of writing specific to orality, are not an imperfect anticipation of cinema, Saussy provocatively suggests, but a culmination of older technologies of writing. I would add that these extensions of writing reappear today on a radically new technological level in digital text-images. Thus Lev Manovich, for example, performs a surprising return to precinematic devices such as painting and animation in order to explain digital cinema, in the process leaping over the Bazinian obsession with photography as the shadow of the real.[2] In both cases, a radical break in *techne* is bridged over by a continuity in *episteme*.

This book's pursuit of the constructions of orality substantiates (I am tempted to say "proves") through a network of historical examples drawn from a rich array of interconnected disciplines—literary studies, anthropology, linguistics, psychology, science, religion—what has until now been articulated only theoretically and therefore, in our theory-hostile age, doomed to obscurity and mistrust: that oral literature (and the oxymoron is a portent) has always been a form of writing, indeed of *arche-writing*, and that the difference between oral and written poetic production is not one of kind, but of *différance*—the active production of a divergence.[3]

The drawing of the boundary lines, Saussy conveys, has been historical and ideological rather than substantive. From time to time moments of rapprochement occurred, as in Caesar's appreciation of the written quality of Druids' orality, or in the invention of orally punctuated *vers libre*. Rhythm is the technology of oral inscription, and the human body, with brain and muscle (including all their varieties of technological extension), has been for ages its material base. Saussy's "show and tell" should be read back to back with Derrida's *Of Grammatology*. It then appears as a tour de force of mediating function, bridging over the painful rift between philosophy and philology, theory and practice, while rendering accessible and crystal clear what seemed to be so cryptic.

The book remediates the philosophy of writing with an ethnography of orality.

This ethnography is twofold. On the one side, it describes many modes of embodiment (for example, hain-teny, Scott's writing machine, the neuropsychology of Ribot and Janet). They demonstrate the corporeal basis of oral literature. Ethnography opens the way to what the new-media theorist Katherine Hayles calls a medium-specific-analysis (MSA), the acute need for which oral literature shares with electronic writing: "Understanding literature as the interplay between form, content, and medium, MSA insists that texts must always be embodied to exist in the world. The materiality of those EMBODIMENTS interacts dynamically with linguistic, rhetorical, and literary practices to create the effects we call literature."[4]

Oral texts are structured by the materiality specific to the human body, just as electronic texts are structured by the materiality specific to the computer's software and hardware. Neither can be read without awareness of the material artifact. Orality has been always based on the system of a complex corporeal apparatus: sound, ear, brain, memory, muscular movement, articulatory organs, sound.

Human bodies were writing machines before writing machines existed. In this sense, oral texts are prototypical "technotexts."

But "textual functions must not only be based on the marks appearing on screen but also [have] to take into account what was happening inside the machinery."[5] This machinery, in the case of oral literature, might take the form of motor-psychology, or of springs, tubes, and metronomes attached to human faces to record their speech.

The discovery of oral literature as another form of writing and another form of materiality is not unlike the discovery of electronic literature as another form of speech described by Hayles: "Writing, a technology invented to preserve speech from temporal decay, here is made to instantiate the very ephemerality it was designed to resist. [The reader] understood that her relation to this writing was being reconfigured to require the same mode of attention she normally gave to speech."[6]

The theory of orality is thus a media theory.

On the other side, this book's ethnography also pursues the customs of academic scholarship. "Oral literature" is a nonexistent subject matter that can be approached only by inference and approximation, always already mediated through some sort of writing. The book retraces decades of the gradual formation of a pattern of thought, bringing across the humble idea that scholarship is always a matter of collective endeavor and that academic writing is always an adventure filled with hazards and contradictions. Scholarship on orality was generated, we see, again and again, as if in successive throws of the dice, out of the code of the Homeric problem, later generated anew in biblical studies, and diffusing its conclusions through a variety of academic disciplines.

Scholarship is a global village, where insights won in Madagascar turn out to apply to the most ancient Chinese poetry. Voices from a distant

African island are echoed in Prague, Paris, and Harvard. In intellectual research, as in art, Saussy teaches us, the slightest differences in the choice of words, the versions of reprinted articles, the accidents of translation matter: they change or reveal the meaning.

Scholarly texts are honored here with the same precision of reading with which literary critics honor literature. No insight is awkward or shallow enough to be discarded. Jean Paulhan set out to study the Merinas but was taught by them. We can't be reminded enough that the path of the intellect is never straight but tortuous. It forks and loops. A fallacious statement, if not taken absolutely, but read next to another fallacy, is a way of discovery.

The book captures the fluid complexity of the intellectual process, without restricting itself to end results and foregone conclusions. In that, maybe, it mimics, in writing, certain features of orality: "the gradual construction of thoughts during speech," as Kleist put it.[7] The reader is treated here to a rare adventure of reliving the process of unbiased research, equipped with a new insight that subtly moves aside the crutches of various ideologies in order to see the pragmatics of the problem for what it is. The multifaceted notion of orality slowly emerges out of a maze of crosshatched strokes, constantly shaken and reconfigured in a kaleidoscope of ideas, under gentle and attentive questioning and dialogue.

As to the refreshing quality of its intellectual surprise, this book gives me the same joy of unexpected constellations of ideas, of unpredictable turns and reversals, as Katherine Hayles's spirited writings on electronic materiality, or Byung-Chul Han's witty techno-political aesthetics.

How to capture the experience of reading this book? To me, it felt like opening a rusty window in a stifling library vault. Imagine that familiar labyrinth of dusty shelves full of old-fashioned, unreadable, methodologically obsolete and awkward books in different crabbed scripts, among which you got lost for years. But suddenly, a blast of wind raises the dust, a beam of sunlight makes visible forgotten names and titles, and you— while looking out into a green, fresh courtyard and breathing in cold, crisp air—hear them talking.

Preface

The dwarf planet Pluto was discovered by indirect evidence. No one saw it through a telescope before its existence had been deduced from the wobble it imparted to the orbits of its nearest neighbors. This has always seemed to me a rare kind of triumph: to see, merely by thinking about it, a thing that had to be there but that couldn't be located.

So here is another book about things that don't speak for themselves, but have to be characterized from the way they affect things around them. Even large and visible objects in the humanities sometimes gain from the indirect approach. I have tried to carry out an investigation of oral tradition, not by starting from a description of oral tradition, but from the observation of the difference it made for people to be talking about oral tradition. What are the conditions of its appearance as a figure of thought? In handling the history of ideas, I have tried to ask, not "Is this idea correct or does this anticipate what we would say later on?" but "How was it possible to have this idea, what supported it, and in what directions could one have gone from there?" Like Rousselot, I set out to walk a dividing line and found it elusive. As the investigation went on, I came to the conclusion that oral tradition was more clearly detectible in the wobble it caused than in anything else: that is, to put it more generally, that a medium manifests itself always and only in relation to other media.

Oral tradition, or "orality," is a matter of literary history and of literary theory. These two subprofessions do not quite coincide in this shared object, because their interest in it differs. Literary history documents how this or that tradition is recorded; theory asks how to think about what generally or potentially happens. The pivotal moment in applying this history theoretically is that at which it becomes necessary to define the properties that a text would have if it were to be considered "oral," absent any knowledge of the events that led to its composition, for at that moment orality becomes a question, not a description, and a question to be resolved by the understanding, not by an appeal to fact. The investigation led me mostly to French sources, where I found a sustained if nearly forgotten development of the question, anchored in many other controversies of the modern era. In choosing to follow this path, I gave in to the temptations of a connected, persuasive, and coherent narrative that brings out with particular clarity the artistry of texts and their relations to publics, media, and authority. This narrative, it seemed to me, would invite comparative rejoinders better than a narrative in another form.

The focus thus achieved brings limits as well. Other histories, defined by different keywords and occupying parts of the same discursive space, could have accounted for the emergence of "orality" too. In the German-speaking world, many of the questions I discuss here would have come into focus through the ideas of rhythm and memory in music, visual art, nature, and the human body; an English-language study would have been centered on the ballad genre and its migrations, rather than on the sustained epic narrative that frames the controversy in France; in China the social disparities between those who sing and those who write would have predominated. (For a kindred history in the Germanic domain, see Hanse, À l'école du rythme; on the British version, see Trumpener, *Bardic Nationalism*; for accounts of Chinese explorations of orality, see Hung, *Going to the People*, and Chen, *Huang tudi*.) I allude to these and other parallel stories along my way.

Although the book's center is in Paris in the years 1660–1960 and its main character Homer, its circumference is wide. Pursuing the thread of oral literature leads us through the development of a *systematic* conception of language and literature—the importance of which for the twentieth century hardly needs demonstrating—and efforts toward a *universal* conception of the same, inherently more debatable. In it can be seen a parallel or anticipatory history of modern critical theory. Well before "the death of the author," the dissolution of that estimable person had been observed by folklorists. What I am calling the ethnography of rhythm gave rise to concepts of intertextuality, generativity, narrative deep structure, the distinction between genotext and phenotext, and the like. In the course of struggling with the alterity of oral traditions, it gave vivid expression to many constitutive tensions of modernity's self-understanding.

I should admit as well to a personal stake in the matter. I find repetition hard to bear—this is one of my failings—and oral poetry is full of repetitions. Forcing myself, years ago, to read every line of Homeric boilerplate as if it had never been said before, submitting to the discipline of learning to enjoy every rosy-fingered dawn as if it were the first, helped me to stand aside from my impatience. But the theory of orality is full of repetitions too, and for these I had less tolerance than for Homer's. Time and again a charismatic teacher pronounces a doctrine, reiterates it in six or twelve successive publications, and attracts a circle of disciples who repeat, their whole career through, the master's theses. I have found that this form of fidelity to the teacher gets in the way of going to the historical origins or the logical consequences of what is said, and accordingly have taken every opportunity I could discover of veering off from strict reproduction. Of course, I must have failed to discover many such opportunities; where I commit an act of involuntary obedience, please take it as homage to the teachers I have heard and the authors I have read.

Figures

le terrain limitrophe où les éléments des deux ordres se combinent ; *cette combinaison produit une forme, non une substance.*

—FERDINAND DE SAUSSURE

Introduction: Weighing Hearsay

The perturbation caused by the idea of oral literature is the subject of this book. It is not a survey, history, or typology of oral literature as such, nor (until the last chapter, anyway) another theory of orality. It does not attempt to set up orality as an answer or a rival to written culture, whether in the name of epistemic relativity, historical preservation, or cultural equality. Nor is it about themes or situations of orality in literature.[1] It examines the history of the concept's formulation and seeks to understand the difficulty of articulating "oral literature."

Part of the difficulty is the implication that by "oral" we must mean "not written"—with strong emphasis on the "not." Much of the *Literatur* about oral literature is devoted to enumerating the properties that distinguish it from written or printed expression. Consider the famous statement of the medievalist Francis Magoun in 1953: "Oral poetry . . . is composed entirely of formulas, large and small, while lettered poetry is never formulaic."[2] This is clear and uncompromising—especially in brief quotation—but almost certainly wrong to the degree that it is clear and uncompromising. It proposes ideal types transcending the endless registers of detail with which Magoun, as paleographer and text editor, was necessarily familiar. Anglo-Saxon narrative verse has come down to us through a chain of manuscript and printed sources, so describing it as "oral literature" must amount to a claim about a form it had prior to the beginning of the chain; or, more tendentiously, to a claim that the features in it that evoke oral literature are somehow more genuine, deep, and essential than other aspects.[3] It means that "oral literature" is what the work *originally* (despite the passage of time) and *really* (despite the words on the page) is. Such claims are necessarily value-laden, interpretive, and contrarian—not a bad sort of claim to be making, but inherently difficult to document.

Both Magoun's assertion of "formula" as a distinctive property of oral literature and his ensuing methodological wager (that counting the percentage of such repeated phrases in a body of text would yield unambiguous results) derived from the work of the American classicist Milman Parry and his student Albert Bates Lord. Magoun's instantaneous clarity ("entirely A, never B") does not hint at Parry's process of discovery. It was only after compiling hundreds of pages of statistical tabulations on epithets in Homer and their metrical correlations that Parry dared (or thought) to raise the question of causality: what might have motivated the epic poet to amass such a vastly redundant yet specialized vocabulary of phrases? Parry's later journeys to record the illiterate epic singers of Yugoslavia were designed to bring ethnographic backing to a hypothesis that had emerged from an argument framed entirely on the basis of texts.[4] But Lord, Parry's assistant during those travels (1933–35), encountered the statistical dominance of "formula" and the spectacle of epic performance at the same time: for him, it seems, they were never distinct.[5] Following Lord, Magoun expresses this identity between the "formulaic" and the "oral" together with its converse, the identity of the "written" and the "nonformulaic."

Established as a duality, "oral vs. written" was now available to the English-speaking world as a cultural diagnostic. Marshall McLuhan saw himself as adopting "the enterprise which Milman Parry undertook with reference to the contrasting *forms* of oral and written poetry" and extending it "to the *forms* of thought and the organization of experience in society and politics." His *The Gutenberg Galaxy* (1962) is a "study of the divergent nature of oral and written social organization."[6] Thus, the theory of orality is presupposed as the original media theory, the basis on which the subsequent "extensions of man" could be thought and accounted for. Divergences come thick and fast in *The Gutenberg Galaxy* and its sequel, *Understanding Media*: hardly any phenomenon of Cold War society, be it Soviet propaganda, Mau-Mau, "the lonely crowd," or the atom bomb, escapes classification as a version of the conflict among oral and written cultures, with radio and television set to wrap the "global village" in a new form of orality. Eric Havelock's derivation of Greek philosophy from the practices of alphabetic writing offers another media theory in which the means of communication shape the possibilities of what can be said and thought, with the twilight of "orality" the moment of crisis and transformation.[7] Less adventurous, but no less committed to the significance of the divide between oral and written stages of civilization, Ian Watt and Jack Goody describe "the consequences of literacy" for the emergence of modernity, settled law, rational thought, and individuality. Societies without writing are "homeostatic": reluctant to admit change and limited in their ability to detect it. "The individual has little perception of the past except in terms of the present; whereas the annals of a literate society cannot but enforce a more objective recognition of the difference between what was and what is."[8] The medium is the *mentalité*. Walter Ong, in a frequently cited passage, draws up a list of the distinctive properties of thought and expression in an oral culture. The oral mind is "additive rather than subordinative," "aggregative rather than analytic," "redundant," "conservative," "close to the human lifeworld," "agonistically toned," "empathetic and participatory rather than objectively distanced," once more "homeostatic," "situational rather than abstract."[9] Ong's "psychodynamics of orality" have much in common with earlier theories of "the primitive mind" which he disavows.[10] The same descriptions lead to the same verdict of an incompatibility between oral and literate that is not one of degree, but of world to world: "There is no way to refute the world of primary orality. All you can do is walk away from it into literacy" (53). Inasmuch as he has selected "the characteristics . . . which are most likely to strike those reared in writing and print cultures as surprising" (36), Ong's profile of orality may be nothing more than a reflected theory of literacy.

Such histories of orality come with ambitions and structures that make them easy targets of critique. They invest so much in the separation of oral from written literature that they are committed to making all oral traditions look very much the same, and different in the same ways from a similarly totalized portrait of written traditions. The sharp divide they draw between oral and written amounts to a "denial of coevalness," in Johannes Fabian's phrase.[11] They are perforce committed to a special form of determinism—the shaping of culture by communications media. But are not these the typical excesses of any emerging field? Without them, the study of media in the 1960s—and the study of orality as a part of it— could not so easily have presented itself as distinct from social history, the history of ideas, or the history of technology. Excessive and grotesque when considered in isolation, the gestures of differentiation between orality and literacy belong to the rituals of argumentative sociality. Academic fields must make their objects visible. To do that, it is sometimes necessary to block the light given off by other, more brilliant and familiar objects, such as (in this case) writing. Research into orality had to eclipse the literate sun—or, to put it less dramatically, it had to pull the shades to shelter the faint light of a microscope pointed at the vestiges of orality.

"Oral theory has become our critical conscience," intoned James P. Holoka in 1973 at the apex of the movement.[12] Not all studies of oral tradition are so ambitious. Another current is the empirical or enumerative style. An example is Ruth Finnegan's *Oral Poetry*, written after the first mad wave of media studies. Finnegan takes it as her mission to "make questionable some of the confident generalizations made about the whole category of oral poetry." Considering one by one the claims of Lord, Ong, McLuhan, Havelock and others, she regularly advises that "in one sense this is true . . . But taken to extremes the approach can be misleading"; "It is easy to exaggerate this . . . and when one comes to the extreme formulations . . . the contrary evidence seems overwhelming"; "one must not labour the point overmuch," and the like. Oral poetry is "not a single and simple thing," rather a "relative and complex term." But other than throwing cold water on overexcited theorists, she has, she says, "no hopeful overall theory to venture myself."[13] Her method is simply "ostensive": she tells you that something is an example of oral poetry, points out a few things about it, and moves on to the next one. The examples are interesting, but they add up to something known in advance: that oral poetry is poetry that circulates by recitation in societies in which writing is not the central means of recording words and deeds. It occurs as event, not as object, in performances that combine memorization and improvisation, usually in a language marked off from everyday speech in rhythm, word choice, or other features. "Orality" has here become no more than a delivery mechanism for the poetry, a fact about its provenance; to ascribe too great a formative power to the delivery mechanism would involve theoretical commitments that the ostensive style of investigation discourages.

Neither the unsustainable gesture of radical separation between orality and literacy nor the low-risk enumeration of examples gives rise to the perturbation that I seek to trace here. Both ways of proceeding—and between them they account for the majority of what is published on oral traditions today—seek to stabilize their object of study, though by different means. To do otherwise, to hold this object as questionable, will require us to forgo an advance knowledge of what "orality" (oral poetry, oral tradition, oral transmission) is and to look into the history of its emergence as an object of discourse. In other words: to treat it as an X and watch what people have written about it, while allowing for the possibility that the means of description will leave something of the object permanently unexpressed.

By calling it "The Ethnography of Rhythm," I have tried to foreground a constitutive distance between orality and its observers, and to insist that it is only through an overlay of different media (since *ethnography* is an overlay of different cultures) that oral poetry emerges into view at all. The book is a working out of the *ethno-* and the *-graphic* sides of the problem. I have tried to warn myself off the naiveté of direct reporting, of telling you what orality *is*.

As has often been pointed out, "oral literature" is a paradoxical name.[14] But it is not necessarily impossible or self-canceling: the "letters" of which literature is (etymologically) composed have changed their nature and functions many times in recorded history, and there is no reason that such a wide field of inscriptions should exclude the human voice.[15] If literature is redefined as "verbal art," oral expression and transmission can be absorbed into it without much ado.

Yet, for most of recorded history "oral literature" went without a definition or a theory. This is not to say that oral tradition had no existence as an object of discourse. It did; but it was a functional concept, invoked only in the roles of contrast or supplement. The "oral" was what was not written; and its meaning, practically, was exhausted by that negation. (For comparison's sake, consider the poverty of a theory of masculinity whose only way of defining men was as "non-women.") Practices of oral transmission were mentioned by writers for whom books and papers were the usual, adequate, and unremarkable method of perpetuating knowledge. What the "oral" is in itself has to be guessed from the assumptions and purposes of the writers as they confront this unfamiliar practice.

Julius Caesar interrupts his account of his campaigns in Gaul in book 6 to offer an ethnographic sketch of the enemy. The higher stratum of Gallic society consists of two groups, the knights and the Druids.

> The lore [*disciplina*] of the Druids is thought to have been transmitted to Gaul from Britain, where it originated. Those who most eagerly wish to acquire it go there for the sake of study. . . . There, they are said to learn by heart a great number of verses, and not a few spend up to twenty years in learning them. Nor is it considered in keeping with divine law to commit these verses to writing, though [the Gauls] use Greek letters for almost all other kinds of public or private business. It seems to me that this rule was established for two reasons: one, that they did not wish this lore to be acquired by the common people, and two, that they did not wish the learners to rely on letters and thus apply themselves less strenuously to memory, as generally happens to those who, through the help of writing, lose their facility

of learning and their memory. The chief thing that their lore teaches is that souls do not die, but rather travel, after death, from one person to another; by this means they consider that warriors can be urged to great feats of courage, as the Gauls have no fear of dying. [In those verses] much about the stars and their movements, about the size of the earth and the different lands, about the roles and powers of the immortal gods, is discussed and transmitted to the young.[16]

The Gauls possess a vast and complex body of knowledge (*disciplina*)— knowledge about the gods, the soul, cosmology, and geography. Many of their youths spend up to twenty years being drilled in the verses in which it is contained. Up to this point, the Gauls' system of education sounds like that of the Roman patricians: a grounding in the poetic corpus constituting what Havelock would call their "tribal encyclopedia," then a step up to more speculative areas of knowledge, finally a period of study abroad. First Homer, then Plato, then Athens. Caesar's imaginary hearer must be impressed by the figure of twenty years for a well-rounded Gaulish education, making the Gauls seem like Indian philosophers. The first word of the next sentence ("*Neque fas esse existimant . . .* ") responds to possible objections to the previous sentence: not only do they spend twenty years being educated (first surprise), but they do not even use writing for their voluminous sacred teachings (second surprise). "*Fas*" suggests a divine or ritual prohibition, as if raising the seriousness of the interdict in response to the hearer's amazement. Caesar, using the first person for the only time in the *Gallic Wars*, then ventures two opinions that demonstrate his command of the Greco-Roman educational corpus, for they are drawn from Plato's *Phaedrus*: the Gauls forbade writing their wisdom down either to keep it from falling into the hands of the undeserving, or to protect their memories from the decay initiated by the use of writing.[17]

Here we have a division of the field of information among the Gauls into two: everyday business, where writing is permitted, and the realm of the sacred, where information is transmitted from person to person over a number of years and in verse form. Oral tradition is marked out as the field of greater value, as writing down the sacred verses would have debased them. And the content of oral tradition in at least one respect mirrors the means of its transmission. Just as the verses memorized pass from one generation to the next, so the soul is said to *"post mortem transire"* to another body, without, therefore, ever really dying. Caesar is unable to say much about the technicalities of Gaulish teaching, but the little he does say inscribes orality in the realm of sacred and immortal things, predicated on an identification of speech with the soul and a rejection of writing.[18] Unlike the failure to write, which could be ascribed to any barbarians, the Druids' refusal to write puts their learning among mysteries that Caesar's page—like any page—cannot contain.

Caesar imagines, in writing, a culture in which writing is not the normative member of the pair "writing/orality." With Flavius Josephus, a century and a half later, it is the more usual reverse. Discussions of epic authorship often quote the passage in which Josephus remarked that Homer "did not leave his poetry in written form; but it was preserved in memory and only later pieced together out of the separate recitations."[19] With these "recitations" (ᾀσμάτα) of a poetry "preserved in memory" (διαμνημονευομένη) we have the ingredients of oral tradition and performance. But Josephus was not offering a scholarly or ethnographic account of the emergence of Greek epic, and least of all was he praising the singer of tales. His purpose in introducing the idea of oral tradition was to discredit it by comparison with a purportedly better-guarded written tradition, that of the Jews:

> However, there is not any writing which the Greeks agree to be genuine among them ancienter than Homer's Poems, who must plainly be confessed later than the siege of Troy; nay, the report goes, that even he did not leave his poems in writing, but that their memory was preserved in songs, and they were put together afterward; and this is the reason of such a number of variations as are found in them. . . . For we have not an innumerable multitude of books among us, disagreeing from and contradicting one another [ἀσυμφώνων καὶ μαχομένων] [as the Greeks have], but only twenty-two books, which contain the records of all the past times; which are justly believed to be divine; and of them five belong to Moses, which contain his laws and the traditions of the origin of mankind till his death . . . and how firmly we have given credit to those books of our own nation, is evident by what we do; for during so many ages as have already passed, no one has been so bold as either to add any thing to them, to take any thing from them, or to make any change in them.[20]

There would be no reason for Josephus to mention oral tradition were it not for the multitude of versions, disagreements, and contradictions imputable to the deplorable lack of writing. With its patchy origins, he implies, Greek historiography comes down to a brawl of specious orators saying whatever will please their various audiences, and that is why a reverently preserved written tradition is superior.

Without Josephus's animus, but using a similar frame of reference, the twelfth-century bishop Eustathius of Thessalonica reported: "That the

Iliad now has a single, continuous, unified, harmonious body is due to grammarians and editors in the service of the Athenian tyrant Pisistratus [r. 560–527 BCE]." Homeric poetry originally "lay scattered in bits and pieces here and there" (σποράδην... κειμένης καὶ κατὰ μέρος διῃρημένης) and was assembled into wholes by different epic singers; these rhapsodes, having no common aim or responsibility, added and took away at will (ὡς ἐβούλοντο). Their depredations ceased only when the authoritative written text was made.[21] The rhapsodic middle period explains whatever in the present text is missing, excessive, or defective. Oral tradition, associated with disorganization, fragmentation, and loss, is a stage the epics have somehow survived: "somehow" because no account is given of the process of collection and editing, what standards were applied, what materials it was based on, and so forth. Eustathius can pass over these details because the workings of an oral tradition, or its transformation into a written tradition, hold no intrinsic interest for him. He writes from within the tradition (single, continuous, unified, harmonized) founded by the written texts; it is these that give him his measure for the earlier condition of fragmentation and variance.

In these last two examples, as in a host of others that might be brought forward, oral tradition is defined by its failure to do what writing is supposed to do. Writing alone gives orality its characteristics and (negative) value; far from preceding writing (as these historical narrations say it does), in the order of reasons oral tradition derives from written tradition. Other profiles for oral tradition are possible. Rather than always doing badly what writing does well, oral tradition can be seen as doing something other than what writing does, or even, exceptionally, as performing the mission of writing better than writing itself can. Plato's fantasy of oral instruction as a "writing on the soul"[22] and its echo in Caesar's *Commentaries* suggest the outlines of one ancient theory of orality: orality as mystery, as transcendence of the physical realm.

But when oral tradition does something other than writing and is explicitly made into writing's complement, its role is not necessarily always the same. The maintenance of oral traditions alongside written ones may have a strategic function in a given civilization: to maintain uniformity while acknowledging difference of opinion, to allow flexibility in practice and continuity in the forms. Oral exegesis changes with the times; the sacred or classic text is meant to outlast them. The Targums expound the written Tanakh, never the inverse. The *hadith* spoken about the prophet Mohammed were "unwritten" only for as long as it took them to be transcribed, catalogued, institutionalized.[23] Oral testimony is indispensable to the modern common-law tradition, but this centrality is ambiguous.

While documents may be introduced into a legal proceeding, their status as evidence requires someone, having pronounced an oath, to vouch for them personally; the need for an oral declaration works to introduce the body of the speaker into a context defined by legal, that is, documentary authority. Speech, however privileged, takes place in a setting of writs, court reporters, and documentation. It does not stay oral for long; it is quickly transcribed, transformed. Indeed, according to John Langbein, the prominence of oral testimony in the common law derives less from a bias in favor of speech than from a recentering of judicial practice on cross-examination in the later eighteenth century. Speech and the body belong to the sphere of things judged, not to that of judging.[24]

In research about the past, knowledge conveyed in speech may be a reservoir, a resource out of which reliable documents can be made: here too is a distinction of potency that modern "oral history" only partly neutralizes. In the print era, archaeologists, historians, and antiquaries might seek the recollections of old people, but did not often take them at face value: hearsay was a plural, protean, "masterless history."[25] It might be validated by comparison with written documents; rarely the inverse. Oral tradition was precisely not a literature; literature was what might survive oral tradition, or emerge from it.

What caused oral tradition to begin to turn into "literature"—to acquire some characteristics in its own right, to emerge from the functional and subordinate role—was a critical dissatisfaction with the written record. Concepts of orality emerge where the master discourse of a written record falters. The seventeenth-century French "Quarrel of the Ancients and the Moderns" shows the two-sidedness of this emergence. At first, the detractors of Homer, seeking to pull him off his pedestal at the summit of literary art, disparage the *Iliad* and *Odyssey* as being "full of weaknesses and violations of common sense."[26] The explanation of the epics' flaws is to be found in the history of their composition, as the abbé d'Aubignac reconstructs it: hymns celebrating the age of heroes, originally performed in princely courts, fell into the public domain and were sung by minstrels and beggars. "Since these verses were common property, always in the mouths of low-class people and beggars, no one valued them, and no learning came to improve this poetry: it was recited everywhere, even on the street corners, which could well have caused the dissipation of the work through negligence, or through contempt."[27]

Nonetheless, some "curious person" gathered up the wandering pieces of these legends and fitted them together, making a new composition that contemporaries called ποίησις Ὁμήρου (*poiêsis homêrou*) or "blind man's poetry."[28] Out of the debris of Homer's reputation emerges a déclassé literary genre, the beggar's monody. The fact that the *Iliad* and *Odyssey* might derive from orally circulated texts is, in d'Aubignac's estimation, a good reason to disdain them, and helps account for their incoherence and poor taste. In Charles Perrault's dialogue on the subject, a character asserts that the very existence of an oral Homer is a charge against the poet, "for if the construction of [the *Iliad*] were, not divine as [Homer's supporters] would have it, but simply tolerable, no one would have invented the explanations I have just presented to you [about a phase of Homeric transmission in the form of disconnected ballads] . . . and it is even more impossible for chance to have made an admirably structured narrative out of a heap of unrelated fragments."[29] Homer is not a sufficiently classical writer to win the approval of the moderns. But this same iconoclast Perrault, stepping outside the frame of classicizing literary polemic, also wrote down the *Mother Goose Tales* as an effort at transcribing authentic folk creativity, thereby becoming analogous to the "curious person" to whom d'Aubignac assigned the assemblage of the *Iliad*.[30]

A self-aware age of "criticism" added the hypothesis of oral transmission to its repertoire of devices for purifying literary taste. The "battle over Homer" (*la querelle d'Homère*) spread through aesthetics, philology, theory of translation, and historiography. Antoine Houdar de la Motte, in his versified adaptation of Anne Dacier's *Iliad*, removed the elements that neoclassical taste found inept, thereby improving (in his terms) the author—an effort many of his contemporaries, including Dacier, found laughable, arrogant, and hypercritical.[31] "Homer never existed," "the composers of the Homeric poems were nobodies": the ease with which either of these arguments modulates into the other in the neoclassical period masks a fluidity in the concept of authorship. A genuine author is *one*: "one" meaning "numerically unique" or "single-minded." Homer should have been in control of every aspect of his artistic design, if he were a genuine author. To show that the poems lack unity of style, argument or effect can mean either that Homer was a mediocre author (Houdar's thesis) or that he was an unruly crowd (d'Aubignac's thesis).

At the same time, in another sense of the word "critical," oral transmission joins forgeries, interpolations, and copying errors as a form of textual corruption.[32] The ambiguity of authorship (is it a numerical property or a degree of intellectual power?) corresponds to an ambiguity of criticism (is it an artistic standard or a scholarly method of discrimination?). Speaking for the scholars, Ludolf Küster echoes d'Aubignac on oral recitation and its murky provenance, turning it to the benefit of text criticism: "With the passage of time, Homer's poetry became thoroughly corrupted by its rhapsodes, many of whom imported their own verses into it. This created the necessity for critics of emending Homer. . . . Critics restore the foremost authors to their original glory, as if polishing the setting [of a jewel]."[33] Although one passes for arrogant and the other for self-effacing, the revisionist translator and the antiquarian have the shared aim of rectifying Homer, that is, to eliminate the traces of the "nobodies" who might have meddled with the poems.

The only role envisioned for oral transmission is that of entropy, of increasing the disorder of literary works. "Critical" awareness, in the era of the Ancients and Moderns, is resolutely set against entropy: which is to say, against oral tradition as then imagined.

Giambattista Vico's "discovery of the true Homer" was a rebuke to the critical consciousness of the seventeenth-century moderns. More powerfully than any other example, Homer allowed Vico to split apart the two senses of "critical," the criticism of taste and the criticism of authenticity. The very characteristics spurned by the moderns in Homer—"his vile sentences, his base customs, his crude comparisons, his awkward words, his metrical license, his inconsistent mixture of dialects, and his making gods of men and men of gods"—proved him to have been a primitive man, a barbarian, "shut up in the body and unable to see outside it."[34] Vico's Homer did not write, and was certainly a "nobody" in the sense that "Homer" was only a general name for the "rhapsodes who went separately about the fairs and festivals of Greece, different ones of them singing different sections of epic poetry."[35] Epic heroes behave like children because they lived in the heroic childhood of humanity, before the invention of letters, law, logic, or even language as we would recognize it; their communication is framed in "hieroglyphs," metaphors, personifications. The subject that speaks in Homer is collective, and therefore alien to us who inhabit a lettered, rational, civilized, individualistic era. Vico's cry of triumph is that "The absurdities and improbabilities that have been held against the Homer known up to now become, in the Homer here discovered, appropriate and necessary."[36] Jean-Jacques Rousseau uses Homer similarly against the standards of both ancients and moderns, making him a version of the wild man whose speech is all passion, figure, and song.[37] Homer's inarticulateness, his multiplicity, his lack of "philosophy"[38] do not corrupt his poetry, but produce it.

Here, then, is a third form of perturbation, the imagination of a profoundly "other" barbarity. When applied to this primitive orality, criticism as understood by the partisans of both the ancients and the moderns is beside the point and will always deface what it purports to improve.

Both the entropic and creative registers of orality are at work, and somewhat obscurely at loggerheads, in Friedrich August Wolf's famous

Prolegomena to Homer (1795). Oral transmission explains the "certain peculiar corruptions and more, and more serious, changes" of the text, as it
did in the traditional accounts of Josephus, Eustathius, and others whom
he quotes at length.[39] But when Wolf states that the original versions of
the poems "were not consigned to writing, but were first made by poets in
their memories and made public in song, then made more widely available by the singing of the rhapsodes,"[40] and cites Ossian, Rousseau, and
Robert Wood for support, he imagines a primary orality as making, rather
than endangering, the poem, or at least as making the many shorter songs
and episodes that would have preceded the *Iliad* and *Odyssey*. This orality
has to be imagined, as it has left no direct traces and its speaker is often
a collective or a national one.[41] Nothing can be said about it that is not
hazy and general. And Wolf's entry into the detail of the poems' history
and the process whereby many short heroic lays were combined into two
huge epics is held back for the second volume, which never appeared, of
his Homer edition.

In another discovery of entropy, the Bible had simultaneously become
a historical document, framed from perspectives and laden with errors
that could not have been ascribed to an omniscient God. Spinoza's *Trac-
tatus theologico-politicus* (1670) set the tone, soon followed by Richard
Simon's *Histoire critique du Vieux Testament* (1678, 1680). Simon thought
the books of the earlier Testament had been shaped out of older, originally archival, materials. The secularizing implications of this proposal
were clear to Bossuet, who intervened with Louis XIV to have the book
censored. (Simon's work was continued in Germany by Johann Gottfried Eichhorn, whose *Einleitung ins Alte Testament* [1783] was a model
for Wolf.)[42] The critical study of the New Testament begins when careful
readers can no longer see the Four Gospels as a single, four-sided, inspired
text, and the discrepancies among the evangelists become irreconcilable
and undeniable. Through these discrepancies, hints of the conflicting interpretations of Jesus and his mission advanced by different factions of
the early Christian movement resurfaced. What had the original Gospel
been? Which parts of the Gospel narratives might be marked off as late
additions and thus inauthentic, though long enshrined in the creeds of the
different churches? Simon, Reimarus, Lessing, and Herder approached
the problem variously.[43] A first expedient was to separate John from the
Synoptics, but this still left many discrepancies, and echoes that were no less
puzzling. Could a sequence of phases of composition, editing, and borrowing
be reconstructed? Lessing surmised in an essay (published posthumously
in 1784) that both discrepancies and similarities could be explained as a
residue of oral traditions that each evangelist drew on in his own way.[44]

Lessing's contemporaries were quick to denounce the potential for abuse, as without written Gospels, the message of Jesus was indefinite and carried no authority.[45] Entropy threatened, in the shape of unlimited theological license.

Herder, in the unusual position of defending the credibility of John, portrayed the three Synoptics in 1797 as "evidently nothing more than variants of one and the same saga that was first transmitted orally hither and thither, finally was written down, and has come to us in three editions."[46] Herder refers to the Synoptics synoptically for the rest of his book *Von Gottes Sohn* in the singular, as "the saga" (*die Sage*). Gieseler revived the oral hypothesis in 1818, holding that the earliest Christians, like the Druids described by Caesar, deliberately avoided the use of writing and recited an oral gospel, in Aramaic at first. To defend against his readers' likely suspicion of the integrity of a gospel so transmitted, Gieseler insisted that the "identity of education, the linguistic poverty and the simplicity of character" of the early Christians ensured that, "as in Homer and the Old Testament, the messenger always repeated the exact form of the message as received."[47] The variability of the gospel and its messages must then date from its confinement to written forms. Already it is clear that the written record does not account for itself; something else is needed that bridges the rifts in Holy Writ; now an empowered oral tradition takes on the status of source and origin of the imperfect and multiple written versions.

In all these instances, orality emerges as a historiographical backformation. No one observes it, it is only inferred. Oral transmission accounts for the insufficiencies of the text as discovered by critical scholarship; oral composition points us back to the imagined plenitude of an epoch of song and recital. As David Friedrich Strauss later put it, romantic theologians

> imagined the first proclaimers of the Gospel on the pattern of the Homeric rhapsodes, in whose mouths the Homeric poems lived on in such a way that they underwent many transformations and additions. Such an analogy was suited to the times, when a deeper understanding of the mentality of antiquity was sought, and likewise a livelier depiction of the origin of poetry and religion. Through the medium of this purely oral transmission, the history of the Gospels appeared as a living thing that could grow, sprout trunks, and send out new shoots and branches.[48]

As a theoretical entity, oral poetry or the oral gospel compensates for the written text and thus clings, logically speaking, to its wreck. By allowing the authority of written records to appear artificial and secondary, "oral tradition" perturbs in one way; by replacing the traditional figure of the author with an anonymous horde or a *völkisch* vision, it perturbs in another way. The invocation of orality is first a critical gesture when performed in d'Aubignac's and Perrault's takedowns of Homer or in Simon's "histoire critique," and then a fictional one as it makes up for the initial destruction by declaring a new "true Homer" placed before all writing, an "original gospel" preceding all churches. The failure of certain texts to live up to their status of literary works is diagnosed as a consequence of illiteracy in the sense of Josephus: orality as an inferior, inconsistent shadow of writing. The desire that the imperfections of the text should be seen as reversible accidents creates for the original recitation, the prototext, a status like that of Caesar's Druid mysteries, an exalted form of writing on the soul. In written culture, orality oscillates between these two statuses.

In Wolf's time there were as yet no books designated as oral literature, unless Ossian counts.[49] With that doubtful exception, the category's home was the historical imagination. The logical possibility of a literary creation that did not involve writing having been established, explorers, historians, dialectologists, and writers of memoirs began to designate such creations with the phrase "oral literature." In the nineteenth century, the term is classificatory when it is not hypothetical. Thus an anonymous "Essay on American Literature" from 1815 catalogues "the oral literature of the aborigines . . . the oral literature of the Indian," asserting that the natives of the continent not only had verbal art but preserved it through the generations.[50] In an analogous circumstance, James Montgomery in 1830 defends the ancient Britons from the charge of barbarity with a reference to their "oral literature."[51] Guillaume de La Landelle (1861) introduces a ghost story as "a study of the oral literature of sailors."[52] Literature where none was expected: such is "oral literature." François-Marie Luzel (1869) describes "the oral and traditional literature of the Bretons" and Paul Sébillot issues a book of folktales under the same label (1881).[53] Sébillot persuades a publisher to use the term as the name of a series, a clear indication that its paradoxical quality has faded from view. But with Émile Chasles (1866), a shade of irony or metaphoricity had already crept into the term, when he described *Don Quixote* as a novel torn between two media: "Oral literature, made from folk proverbs, and written literature, adorned with aristocratic gallantry, intertwine and struggle with one another throughout this novel: it is a battle between the romance and the proverb."[54]

Though the phrase had by 1881 become familiar (in French and English, at least), its theoretical ambitions at the time are modest. Authors implied no more by the term than Ruth Finnegan did. The word "oral" was simply a tag indicating a kind of provenance that some texts might have. It designated the occasion of collecting, but did not say much about the character of the text (apart from the implications of terms like "un-lettered," which applied more properly to their authors). The use of the term "oral literature" at this stage cannot be called a theory; it is barely a description. The person who proposes that cowboy poetry is poetry composed by cowboys has not begun to theorize. But early in the twentieth century, things began to change, with the emergence of the idea (implicit in Vico, but burdened there by his theory of barbarity) that oral texts might be structured differently from literate texts. Or—to acknowledge the weight of tradition—the novelty of the idea was first that oral texts might have a structure at all, rather than being defined by the absence of the kinds of structure familiar from texts originating as written works. A quiver of doubt—the disturbance of orality—begins in Madagascar and is felt in ancient China, in Galilee, in Homeric Greece, in the Slavic domain, and many other places besides.

1 / Poetry Without Poems or Poets

Avis aux non-communistes: tout est commun, même Dieu.
—BAUDELAIRE, *MON COEUR MIS À NU*

After ten years, I no longer saw the ox.
—COOK DING, IN *ZHUANGZI*, "YANG SHENG"

In 1908–10, while serving as a lycée teacher in Antananarivo, Madagascar, Jean Paulhan often observed the sort of scene he described as follows:

> After the evening meal, the children spread a clean mat over the floor, and a group of village men who have been waiting in the courtyard are admitted. They sit down on the mat next to the householders. One of the men opens the session by reciting a few verses. He pronounces them with a forceful rhythm and with such energy that he seems to be voicing a complaint or in some way demanding his due. And then one of the inhabitants of the house, the father or a son or sometimes one of the women, will answer him in the same tones— sometimes brusquely, sometimes ironically. The discussion continues. The audience now and again takes part, interjecting a few rhythmic words that seem meant to redirect the discussion towards its real object. Bit by bit the speeches of the two opponents become longer, more forcefully accentuated; by now each speaker has acquired a cheering section to encourage him with their bravos and their laughter. At the end the opponents are shouting, until suddenly one of them finds the decisive words—or so one discovers when the other hesitates and finds no answer; that speaker then acknowledges defeat and the crowd rushes to congratulate the winner.[1]

What Paulhan witnessed night after night and attempted to record in his notebooks was "a poetry of dispute," performed in the course of "poetic duels." The inhabitants of Madagascar called this poetry *hain-teny*—a term of uncertain derivation, but usually interpreted to mean "the science of language, the knowledge of words"; or in another rendering, "examples in words, exemplary words."[2] Fascinated by the allusiveness and rhetorical versatility of this poetry as well as by its omnipresence in everyday life, Paulhan collected some eight hundred examples and published a selection of these together with an account of the performance practices surrounding them as his first book.[3]

"Two or Three Hundred Rhythmic Phrases"

"To the European ignorant of the native tongue," says Paulhan, "the recitation of hain-teny has the appearance of a bitter conflict of interests, where every hain-teny poem is an argument." From the expression of conflict emerge "a few verses"; a few more verses answer the first; the alternation of points of view continues and a poem is built up through argument and counterargument until one speaker concedes defeat. Most frequently, hain-teny "are uttered for the pure sake of play," but there has to be at least the pretext of a disagreement even when the object of the duel is merely to test the wits of the participants. And when a disagreement really exists,

> it sometimes happens that the visitor who appears as a participant in the evening discussion has had, during the daytime, a conflict or dispute with one of the householders. Improvising in hain-teny is thus for him a means of demonstrating the justice of his cause. One day, in the house of Ambatomanga where I was a guest, a roofer came to call. That day he had finished a job for which he demanded sixty centimes, and Ambatomanga paid him only thirty . . . The roofer who demanded an unusually high salary became a mistreated young girl who had broken with her boyfriend, and the householder, becoming that boyfriend, pleaded with her to come back . . . The householder lost the match, and had to pay the workman at the rate demanded.[4]

Lovers in the place of employer and laborer, poetry in the place of lawsuits: such a transposition of themes might seem a pretext, designed to mask whatever real-life interests are at stake in a bout of hain-teny composition. But this kind of poetry has as its necessary precondition a nucleus of conflict. Poetry does not replace conflict but formalizes and articulates it.

> Near Alasora, two young cowherds are playing with captive crickets. One of them leaves for a moment, and the other steals his crickets.

When the first returns, he sits down, without a trace of anger, opposite the thief and recites a hain-teny. The other answers in the same
style. The argument continues for a long time, without anyone raising
his voice. . . .

I could cite many other incidents of the same kind, where the
discussion in hain-teny derives from a real dispute, which it pro-
longs and resolves. But these are exceptional cases, and it would be
mistaken to think of hain-teny as playing a judicial role in social life.
Normally they are uttered as mere play, and if their recitation bears
the outward marks of a quarrel, it may be artificial.

At Ambatomena, a village of twenty or thirty houses not far from
Tsinjoarivo, a fifty-year-old Merina named Rakotobe debates in
hain-teny every evening with his two sisters, Razay and Rasoa, as
an amusement. The audience is made up of several village children.
Before beginning, Rakotobe sketches out for them the argumentative
pretext that he has made up: one of his slaves has escaped, and his
sister Razay agrees to speak on the guilty party's behalf. On another
evening, it may be a love-rival who has crossed Rakotobe's path, or a
wizard who has put a spell on the family's rice supply.[5]

Hain-teny poetry is formed in dispute and dialogue. It cannot be elicited otherwise, as Paulhan quickly learned when he began to go on collecting tours. "Do you know any hain-teny?" he would ask the village
wise man, and the elder would happily comply, chanting out a couplet or
two—and then fall silent. "Now the men sitting around us would turn to
me and say, 'Answer him, answer him! He won't say a thing if you leave
him alone; he won't have anything more to say!'"[6] Until the second punch
is thrown, there is no fight. Like chess or tennis, the art of hain-teny exists, can be elicited, only in a climate of real or imaginary antagonism
between participants; and to get anything at all out of his informants,
Paulhan had to learn a good many standard responses and join in the
game, at however low a level.

Hain-teny poetry, then, is what Charles Taylor might call an "irreducibly social good."[7] It cannot be created or enjoyed by a solitary subject.
Beyond the formal and semantic characteristics of the individual lines,
what shapes the genre and every work in the genre is an antagonistic
rhythm of call and response, challenge and answer, back and forth, each
verse attempting to crush the precedent verse.

A hain-teny poem has no value by itself, taken singly. Whether its
recitation contributes to a debate with a practical origin and purpose,
or whether it is a mere amusement, it presupposes a real or imaginary

rivalry and hostility that must end with the victory of one of the rivals. Hain-teny is, if I can suggest this term, a poetry of authority. It belongs to a contest of language, where it is nothing but an argument. Its argumentative character is so ingrained that the Malagasy idiom is not to "utter" or "recite" hain-teny, but to "fight with proverbs, to set hain-teny fighting."[8]

Though produced by antagonism, the poetry does more than express disagreement. Directing each participant to stage-manage antagonism is, in fact, one of the refinements hain-teny brings to conflict. Every stanza of hain-teny is, as Paulhan puts it, "an image of the total struggle." "As the discussion follows its course, the man and the woman alternately emerge as the winner: for each reciter gives victory, after a simulacrum of combat, to the combatant he has chosen to represent. . . . So, the better to prepare for his final triumph, [each reciter] first affirms his victory in little struggles over details where he takes on himself the role of the two adversaries."[9] When A performs his "inning" of poetry, so to speak, the rules demand that he represent first his own position and then the opponent's position, giving himself, of course, the better lines and representing his opponent's answers as sensibly weaker. I don't just say what I say and leave you to answer; rather, I say, "When I say such-and-such, you will no doubt answer that . . . " The loser in a debate has not only fallen silent, he has allowed the winner to have the last word in the loser's name. Thus, there is a combination of open-endedness and closure in each episode of hain-teny. I utter my piece strategically, in the expectation that you will answer it, but if you don't, my tentative ending becomes your ending too, and then the ending.

Thus far, then, Paulhan's observations on the performance conventions of the hain-teny. But what are the means of the contest? What exactly enables the roofer to defeat his employer? What brings victory in a combat of hain-teny is not the unprecedented insight, the new angle or the striking coinage, but rather the strategic marshaling and use of stock proverbs known to everyone. The hain-teny is a combat of and by clichés. It is devoid of innovation, of originality, of the verbal "making" that is at the root of poetry's Greek name. A reciter has to pick the right proverbs for the aggressive part of his entry, then provide the right inadequate proverbs for the second part; the person who replies needs to find proverbs that outdo the proverbs of the first speaker, either by appealing to loftier principles or by adhering more closely to the facts of the (fictitious-factual) case. Any overstatement, ambiguity, or inaccuracy is sure to be exploited in the adversary's response. It is significant that so much of the hain-teny situation is conventional. The framework of stereotyped love stories known to everyone in advance bring to the fore the strategic and rhythmic element, the question of timing: both parties responding to each other's stories, saving their best hits for last while trying to get the adversary to squander the whole arsenal prematurely. And it is being able to follow the good and bad shots as well as the depletion of the stocks that makes for the interest of a hain-teny combat. The adept speaker possesses a stock of basic proverbs and the skill to adapt them in new ways (these adaptations may, in the course of time, become proverbs in their turn) together with an experienced campaigner's knowledge of the terrain, the enemy's strong and weak points, the most likely thrusts and the most effective parries.

Lest it seem that the choice of good verses is the thing to aim for, let us remember what Paulhan only slowly came to understand: that what one acclaims in the successful hain-teny is not at all a matter of the word or of the line for that matter but of its place in the context of a dispute. A "good" proverb was not enough by itself to end an argument; Paulhan had before him examples where "slight" sayings won out over "weighty" ones, on account of their more shrewd deployment.[10] The words are there for naught; the same phrases, put in the wrong place, would have no effect or a negative effect on the player's score. Likewise, when Paulhan tried to organize his collection of hain-teny by themes, he found that

> such distinctions run the risk of remaining incomplete and false, if one expects to see in the theme a subject given once and for all, which every reciter develops in his or her fashion. Rather, it must be thought that a hain-teny, in reality, is never alone: one part of its meaning certainly derives from the verses and proverbs it contains, but another part is imposed by the hain-teny it is meant to answer, and the following hain-teny will in its turn pin down the precise implications of the former. Take a poem that seems to express simple pride: if used to reply to a modest, timid declaration of love, it will signify refusal. Another poem ostensibly meant to give advice may [depending on its placement in the series] become acquiescence or mockery. A complex of meaning builds and unbuilds itself moment by moment through a thousand exchanges.[11]

Victory in hain-teny combat rests on memory and analogy, not on the creation of new verses. The adept player will prefer an effective precedent over a novel reasoning. New verses, when they appear, sound much like existing verses, and are often uttered immediately after their models, making the resemblance obvious. A frequent poetic procedure is the elementary assonance provided by the repetition of a single word or word-group . . . [sometimes] not only word-repetition, but repetition of a whole sentence, not so much in the sound and form of its words as in its logical and semantic structure. We find parallelism and symmetry in two, three, four, sometimes even twelve successive verses.

One might imagine a language consisting of two or three hundred rhythmic phrases and four or five hundred verse-types, fixed once and for all and passed on without modification by oral tradition. Poetic invention would then consist of taking these verses as models and fashioning new verses in their image, verses having the same form, rhythm, structure, and, so far as possible, the same meaning. Such a language would quite closely resemble the language of Malagasy poetry: its type-verses are proverbs, and its poems, imagined in imitation of these proverbs, reproducing them in hundreds of new copies, stretching them out or shortening them, setting them for the sake of contrast amid other differently rhythmed phrases, are the hain-teny.[12]

Singular verses, well-constructed poems, original themes, incomparable style, distinctive authorial personalities—none of these are to be sought in this tradition. Yet, Paulhan is sure that the Malagasy hain-teny are poetry.[13]

As the outline of a poetic language, however exotic or theoretical, this is extremely original for its time, precisely because originality is the least of its concerns. In the understanding of the French reading public of 1913, to claim the mantle of poet commits the writer to an avoidance of everything that was proverb, *locus communis*, ready-made thought or stock phrasing. The distinctive sign of poetry was verbal individuality. Baudelaire had claimed, only half-jokingly, to be able to teach anyone "how to achieve, through a determinate series of efforts, a proportional degree of originality": it was all a matter of "the coupling of this or that noun to this or that adjective, analogous or contrary."[14] Mallarmé sought to make a decisive separation in kind between poetry and facile, saleable reportage. The verse, a newly forged semantic unit capable of conferring novelty on its familiar components, is the place where this separation occurs.

> Narrer, enseigner, même décrire . . . l'emploi élémentaire du discours dessert l'universel *reportage* dont, la littérature exceptée, participe tout entre les genres d'écrits contemporains. . . . Le vers qui de plusieurs vocables refait un mot total, neuf, étranger à la langue et comme incantatoire, achève cet isolement de la parole . . . et vous

cause cette surprise de n'avoir ouï jamais tel fragment ordinaire d'élocution.[15]

Narration, edification, even description . . . an elementary use of speech suits the needs of that universal *reportage* to which belong, literature only excepted, all contemporary species of writing. . . . Verse which refashions several units of speech into a new, total word, foreign to its language and more or less incantatory, accomplishes this isolation of the word . . . and surprises you with the impression of never having heard before this or that ordinary fragment of elocution.

A new verse makes speech (an ordinary thing, when used ordinarily) seem something "never heard before." The newness experienced by the reader of poetry pairs off antithetically with the term *reportage*, italicized by Mallarmé as if to make us stop and look at it more closely than usual: *re-portage*, what "carries or transmits again." Poetry, on the other hand, is neither a *re-* nor a *-port*, neither a repetition nor a vehicle. Such a standard was demanding for readers and writers alike. A steady and sterile debate in French letters opposed the healthy banality of common sense and the "children of Mallarmé." As two young champions of "good sense," "nature," or "Frenchness" (the frequently invoked remedies to the excesses of Symbolism) put it:

> [In the era of Symbolism] we saw intelligent young men disregarding the most elementary notions of contemporary science and philosophy and priding themselves on an illusory erudition, solely occupied as they were with digesting the cerebral fantasies, the intellectual mishaps, the superstitious ravings, the literary lunacies, of every cast-off and anomaly that human madness has produced over the thirty centuries since human expression began. (Maurice Leblond, *L'Aurore*, December 26, 1902.)
>
> Under the name of poetry a multitude of imperfect emotions, truncated states of soul, fleeting illuminations, bizarre and unformed thoughts descended upon the world, disturbing everything, alienating good taste, going wide of common sense and threatening to drown French clarity in a starless night. (Paul Souchon, *La Plume*, July 15, 1902.)[16]

The time when Jean Paulhan was investigating the Merina poetry of proverbs and writing up the results fell among the years (1892–1915) in which Paul Valéry, after a period as Mallarmé's devoted follower, aban-

doned poetry. After reading Mallarmé, Valéry recalled, one found everything else "naïve and slack"; the world of letters split into two parts, the
books Mallarmé had written and those books Mallarmé had declined to
write. "This peerless"—or (accentuating the temporal connotations of
Valéry's wording)—"this irreproducible life's work, *cette oeuvre sans se-
conde*, instantly targeted and demolished that basic convention of ordinary language, [which is:] You would not read me if you had not already
understood me."[17] Those who followed Mallarmé tried their hardest to
write verses that no one could understand before having read them. And
this newness, or difficulty, arose from combinations. "In consequence,
Syntax, which is calculation, regained the dignity of a Muse" for Mallarmé's select band of readers.[18] When Mallarmé "created a poetic language almost completely his own by his refined choice of words and the
singular turns of phrase invented or developed by him, refusing at every
moment the immediate solution that the common spirit whispered into
his ear. . . . That was nothing but a resistance to automatism, down to
the details and elementary functioning of his mental life."[19] Inspired by
Mallarmé's example and applying it even more strictly than the Master
himself, Valéry took the refusal of automatism and chance events so far
that he renounced poetry altogether for two decades.[20]

If Mallarmé's definition is what the more adventurous minds in French
letters had come to accept as stating the nature of poetry—poetry as a
problem, possibly indeed, as it was for Valéry during twenty years, an impossible task—Paulhan's ethnographic description offers a different and
in many ways opposite version of poetry and poetic value. What Paulhan
discovered was a strict poetics of "ordinary language," not the rejection
but the espousal of automatism. Not only do the poets of Madagascar
say what has already been said millions of times before, but their artifice,
their skill, instead of being revealed in the "making of new, total, strange
words" out of the preexisting words, resides in their adroit selection and
repetition of whole preexisting verses or varying them in minimal ways.
They do nothing that is "singular" or "completely [their] own" because
their art is common, public, produced by and appealing to the authority
of consensus. They are engaged in a different type of "differential text,"[21]
an "ergodic literature."[22] Making a new verse might even be an infraction
of the rules. If Mallarmé as framed by Valéry's reverent memory is the
extreme case of the poet, the poetics of the hain-teny must be the most
antipoetic poetics imaginable: it pushes to the margins everything that
mattered most in the Symbolist understanding of the art, and makes central what mattered least.

But with this opposition of syntax and originality to formula and proverb, are we not merely making a distinction among things that, though operating on different levels, work in similar ways? Is the combination of old proverbs not a new creation? Most of Mallarmé's words could be found in the dictionary and the daily paper. The difference is that hain-teny revealed a different possible structure for the literary game as a whole. "A hain-teny poem has no value by itself, taken singly"; "a hain-teny, in reality, is never alone": what Paulhan discovered in Madagascar was a social poetry in which neither verse, nor poem, nor poet counted for much individually. The choice of verses had value in relation to the agonistic situation; the stanza rebutted or provoked another stanza; the reciter needed another reciter to answer him. The practices of hain-teny composition, and likewise the artifacts it generates, attest to a logic of reciprocity, of give and take. If Mallarmé believed that the purpose of poetic craft was to "achieve the isolation of the word" from ordinary language, his Malagasy counterparts vied with one another to achieve the maximum imbrication of their acts of speech in a traditional idiom of lexicalized clichés—proverbs—for on that depended the effect of authority they sought.

The proverb's authority is matchless, for the proverb is the repository of common sense, the defining standard of the intellectual apparatus that all competent members of Merina society have in common: if you are not a fool, you will know that a stitch in time saves nine, that too many cooks spoil the soup, and so on. You will also know that those proverbial statements are not often called on to say something about sewing or soup. Thus the proverb's component words practically disappear: its meaning and their meaning have little in common. Proverbs form a "sacred language," as Paulhan termed it when he made a presentation to the Collège de Sociologie animated by Georges Bataille and Roger Caillois in the 1930s.[23] In that talk, Paulhan recounts the misadventures that followed from his attempt to treat the words and images of a proverb as independently meaningful. An elderly patriarch, finding that his sons were set on doing something against his best advice, shrugged his shoulders and said to Paulhan: "What do you expect? The dead ox doesn't chase away the flies." Paulhan objected: "But you're still an ox in excellent health!" It was as if no one had heard Paulhan's little corrective; the conversation went on without him. Afterwards, one of Paulhan's friends went to see him and asked indignantly how he could have called that respectable old man an ox.[24] It is as if the proverb the old man had cited no longer contained the word "ox"; to the properly trained hearer, the ox was invisible and inaudible. But Paulhan had released the animal by breaking the proverb into its component words.

Through his mishaps, Paulhan discovered that the proverb had, in effect, a global meaning as a whole but no lexical meaning as a series of articulated signs. Training in "authoritative speech" could dispense with the details of oxen and flies: it would deal principally in two kinds of information, ten- or fifteen-syllable "words" and the kinds of situation to which the right mega-word would be a conclusive response. Such utterances are sacred, special, immaterial in relation to ordinary, parsed, contingent language. To get to the level where the mega-words and the mega-situations can be perceived, you have to lose track of the linguistic level on which Baudelaire or Mallarmé operated. Skill in everyday Malagasy conversation may even be an impediment. Paulhan's slow education in the protocols of hain-teny could be described as a comedy of the commons:[25] trained to recognize poetry in the singular, individual utterance, the ethnographer gradually comes to find it in the apt recycling of banalities and irrelevancies.

The basic competence of whoever would practice hain-teny consists in "two or three hundred rhythmic phrases and four or five hundred verse-types, fixed once and for all and passed on without modification by oral tradition." Despite the mention of "oral tradition," the fact that the poets of Madagascar are mostly illiterate does not concern Paulhan. He is not in the least interested in classifying their poetics as "oral" or contrasting it with an advanced literate poetics.[26] The important thing about "oral tradition," in his account, is the fact of its being anonymous and collective. Individuals exist in this poetic tradition, for there are winners and losers, and a clever move will result in its speaker being considered clever, but the participants in hain-teny are submerged in collectivity, both by the argumentative situation that gives rise to the poetry and the finite materials from which the poems are drawn. Paulhan sketches a generative model for the hain-teny genre, a combinatory scheme for the production of utterances. Given an overall persuasive purpose, a database of potential verses, and a specific challenge from the antagonist, which as yet unused item from the reservoir will most effectively silence the adversary and add to the speaker's authority? Poets come to their lines by memory and strategy, not expression. They and their spectators demonstrate "unoriginal genius."[27]

Vico and Wolf had already profoundly modified the idea of authorship by making Homer not an individual writer, but the personification of "the genius of Greece." But the passage from the individual to the collective was mysterious, usually papered over with exclamation points. Witness Dugas-Montbel in 1830: "Vico was the first to notice that [epic] was not a work of literature, it was the poetry of an era, the voice of a whole people. . . . Vico opened the age of true [Homeric] criticism."[28] Or Renan in 1890:

> The most sublime works are those that humanity has produced collectively, with no proper name being attached to them. The most beautiful things are anonymous. . . . Do you think you do honor to this or that national epic because you have discovered the name of the feeble individual who composed it? . . . That very name is a lie: not he, but the nation—humanity working at a specific point of time and space—is its true author. . . . The only name that should designate the author of these spontaneous works is that of the nation in which they blossomed; and that name, rather than appearing below the title, is inscribed on every page. Even if Homer were a real and unique individual, it would still be absurd to call him the author of the *Iliad*: the idea that a composition of that sort had issued ready-made from an individual brain, with no antecedents in the tradition, would be insipid and impossible; one might as well suppose that Matthew, Mark, Luke and John invented Jesus.[29]

Paulhan's generative model (prefabricated verses known to all, waiting to be called out for a use on a particular occasion) provided a place for the commons without foisting group subjecthood on the collectivity. It thereby renewed the discussion of oral tradition. For the first time since Vico, a mechanism was proposed whereby a collective subject could become the "author" of traditional texts. The crystallization of this model of text-generation into what we know as the theory of oral poetry can be traced simply by following the traces of Paulhan's book in subsequent scholarship. For Paulhan's study of Madagascarian poetry sparked a rethinking of the ancient Chinese verse collection *The Book of Songs* in a 1919 book by Marcel Granet, was received as a voice of confirmation by the unusual scholar of folk traditions Marcel Jousse, pointed out new directions to the American Homer scholar Milman Parry who is usually credited with the rediscovery of oral poetry in the twentieth century, and assisted the linguist Roman Jakobson and the folklorist Pyotr Bogatyrev in the shaping of Prague structuralism. The pivotal position mapped out by this set of citations is entirely deserved.

Festivals of Rhythm

Apart from brief reviews, Paulhan's 1913 ethnographic account of a social poetry next surfaces in 1919 in Marcel Granet's first major work, *Fêtes et chansons anciennes de la Chine*.[30]

Festivals and Songs of Ancient China is a reinterpretation of one of the Chinese Classics, the early anthology of folk and ritual poetry known as the *Shijing* or *Book of Songs*. Scholars since antiquity have observed that the meter, rhyme, stanza, imagery, and themes of most *Shijing* poems order their content into symmetrical, paired units.[31] To take a typical stanza:

O, the magpie	維鵲有巢
Has a nest;	
And O, the dove	維鳩居之
Dwells in it.	
This girl	之子于歸
Goes to wed;	
A hundred carriages	百兩御之
Escort her.[32]	

Thus the *magpie* is set off against her *nest* at one level (the first pairing of metrical units: two syllables against two syllables, noun phrase against verb phrase) and against the *dove* at another level (the magpie displaced by the dove in the action of the poem; prosodically, one line of four syllables answered by another line of four syllables). The *description of animal life* in the first pair of lines is then balanced by a *description of human life* in the second pair of lines, with a suggestion that the marrying girl is similar to the dove in that she will dominate the house she comes to live in. From the level of the phrase to the overall structure of the poem, the poetic workmanship of the authors of the *Shijing* proves to be based on opposition and complementarity.

The odes of the *Shijing* had always been considered a record of the songs of the people, reflecting their feelings about the virtuous or vicious sovereigns who ruled them.[33] Granet sensed a discrepancy between the lapidary poems and the historical prolixity of the commentaries that had long been attached to them. Rather than commemorating individual kings and queens or recording the details of ancient scandals, the poems seemed to Granet to have a far more abstract and general frame of reference, reflecting persistent social patterns rather than singular historical events. Recent ethnographies allowed him to imagine a social order that would account for both the form and the themes of the poetry. In this early stage of Chinese society, as Granet conceived it, men and women kept apart as a rule and people of different clans met only rarely; perhaps the population was even divided, like some North American tribes, into antagonistic moieties. The festivals of spring and fall gave men and women of different clans the chance to come together. As they did so, they sang to each other in antiphonal choruses or competed for one another's attention through "poetic jousts," the pattern of which Granet derived in comparatist fashion from Paulhan.

> The imagistic formulas . . . originated in amorous jousting: this fact can help to reveal the meaning of those proverb-contests of which the *Shijing* offers an example—a fairly obscure one [i.e. "Xing lu," *Shijing* 17]. In such a contest, each of the adversaries tries to establish a correspondence between the point that he hopes to drive home and a series of conventionally accepted proverbs that no one could deny without being suspected of disrespect for tradition. He draws up his argument in what I call an *analogical rhythm* and buries his adversary under a torrent of ancient formulas: whoever runs out of formulas, finds his traditional knowledge running thin, or can no longer find in the popular stock a workable metaphor or a supporting statement, that one is the loser.[34]

The echo of Paulhan's Madagascar is obvious. But one effect of transregional comparison is notable as well: the example is simplified, shorn of its detail and ambiguities, made to fit cozily in the envelope of a thesis. Paulhan's interest in the game-aspect of hain-teny, the strategies and dodges by which players can turn a stock poetic language into *their* language, the means of their desires, barely attracts Granet's attention, focused as he is on the description of a collective medium—taking "collective" as the denial of personality. This sociological reading of a literary text is insistently privative, as if the purpose were to strip literary reading of its pretensions to individuality, to style, to singular perceptions. Through negations, the contrast with some other kind of poetry—individualistic, expressive, irregular—is kept in the foreground of implication.

> One striking fact about these ancient songs is that *no* personal feeling ever appears in them: not that personal poetry is absent from the *Shijing*, but personal poetry is *not* the inspiration behind the poems examined in this study. All the lovers in this collection are identical: they all express their feelings in the same way. *No* portrayal of a recognizable individual occurs: it is almost always enough to designate the other with a pronoun, the word "lord," or ready-made expressions like "the beauty," "the pure girl," "the wise virgin," "the one in my thoughts," "the fair lady" . . .
> These lovers *without personality* express *nothing but* impersonal sentiments. In truth, what we find in these songs is, rather than sentiments, *sentimental themes*. Meetings, vows, tiffs, separations: in these

common situations every man and woman feels similarly; *no heart* is filled with a unique emotion; *no particular case*; *no one* loves or suffers in an original fashion. All individuality is *absent . . .*

Even the décor is as monotonous as possible. It is supplied by *rustic themes*: the details are concrete, to be sure, but these are ready-made themes and formulas introduced into the song. They make up a kind of obligatory landscape and, if they are related to the feelings expressed, the aim is *not* to make the feelings specific but rather, as we have seen, to link them to general patterns.

The individual case matters *not at all*: that is why these songs can borrow verses or whole stanzas from one another; that also explains how easy it must have been to superimpose this or that meaning willfully onto each song; but above all, this shows that there is *no point* in seeking the personality of an author in any of these poems. . . . The impersonality of this poetry requires us to ascribe to it an impersonal origin.[35]

In the European tradition, romantic love and individualistic poetry are closely associated. The Chinese tradition as read by Granet forms an antitype to this doubled romanticism.

By the very fact that sexual union was, originally and essentially, a means of social cohesion, sexuality could not but be regulated. . . . I see a proof of this strict regulation in the fact that love was unrelated to the fantasies of desire and the caprices of passion. Indeed, in these improvised songs, love always has an impersonal tone: it expresses itself, not through the free play of original inspiration, but through formulas or proverbs more suited to voice the habitual feelings of a community than the singular emotions of individuals. When in the course of jousts, in the heat of competition, protagonists stepped forth and challenged each other to improvise face to face, the source of their invention was not the particular depths of their souls, the characteristic movements of their hearts, the fantasy of their genius; rather, invention arose on the pattern of traditional themes, based on a dance-rhythm followed by all, and propelled by collective emotions. And it was in proverbs that they declared the blossoming of their love.[36]

If there is little free play in the deployment of stock phrases for Granet's rustic poets, there is just as little room for creativity or misunderstanding between the world evoked in their language and the cosmology they shared: "their art is completely primitive. It predates even the use of metaphor. The connections of ideas are as unartificial as possible; they grow from a natural kinship revealed by symmetry, that most basic of artistic procedures. . . . Correspondences existing in reality make themselves known in poems."[37] And poems based on paired words, paired concepts, paired proverbs, end, as if in a magical consequence of this verbal form, by making pairs of their speakers:

> An effect of mutual enchantment resulted, eventually, from the twin litanies of [the male and female competitors in poetic debates]: all they had to do was to sing to each other, alternately, the old formulas of love, which needed only to be repeated, with scarcely a nuance added to the verses: their improvisation was wholly traditional. The inventive genius of these young people was not led on by the originality of their feelings or by their chosen love-object; there was no urge to form a new precedent with novel arguments; the case that concerned them was old and certain, with its repartees established long in advance. To achieve the aim of their courtship, the young people had only to play their parts in the joust; if ever they improvised, their inventions were nothing more than consequences of the dance rhythm.[38]

Paulhan had discovered in hain-teny a social poetry, but Granet makes it a *merely* social poetry. To read the poetry as voices of lost individuals addressing us across the abyss of centuries would be a category error. No such individuals could have existed. This is an account of "the elementary forms of the poetic life." The set attitudes of typecast protagonists—the wise virgin, the brave youth, the timid lover, and so on—that we find in this poetry are not merely expressive devices that a speaker might have used; the attitudes are the actual speakers of an impersonal poetry. The ancient Chinese Odes according to Granet are a poetry of and for all, executed in preset formulas by speakers who do not know themselves as individuals but only as voices of a collectivity, accurately enacting their predetermined roles.

Granet's reading announces a victory of sociology over the personality and originality that had been, for hundreds of years, the special boast and claim of the literary humanities. At any rate, ancient Chinese poetry, in this version, opposes no resistance to the thesis that ideas and texts are ultimately the shadow of "collective representations" society makes of itself.[39]

To declare one's love with a proverb amounts, as we see, to declaring oneself a formulaic unit, a ready-made signifier ("All the lovers . . . are identical"). And just as individuality correlates, in our most habitual vocabulary, with interiority, so the lovers who speak in impersonal formulas perform not so much poetry as dance, the exteriorization and collectivization of rhythm: "a dance rhythm followed by all."

> The feelings to which the jousts gave a versified expression were likewise, by the effect of this mimetic tournament, presented in images. Intense and collective, impersonal and complex, immediate, prior to any analysis, utterly concrete, simple movements of the soul, these feelings found no more apt expression than in the moving figures described by the alternating choruses. . . .
>   It is beyond doubt that, in the poetry of the *Shijing*, all sorts of impressions were associated with the sounds of words and particularly with impressions of movement. How else to explain this but by the fact that gestures came forth to complete the singers' voices, and that pantomime displayed for the eyes what the songs depicted orally?[40]

What the poems of the *Shijing* have to say reduces, for the essential, to the formulas of male and female role-playing in the dance and in the joust, and this reduces yet again to sheer alternation: Yin and Yang, complementary positions, rhythm. Rhythm does not simply organize content; it becomes the content.

> The theme was a given, the tune was familiar; in performance there must have been some degree of improvisation; the fixed verses imposed by them were answered in various ways, so that new stanzas emerged.
>   How did these inventions come about? What was the source of their inspiration? Most probably, it sprang from the rhythm. In these simple poems based almost entirely on symmetry as a mode of expression, rhythm, clearly, was all.[41]

Rhythm, the part of poetry that is least tied to words, representations, ideas or meanings, "was all": and we see how far this "all" can go, as a duple rhythm is the axis on which the construction of stanzas, the formation of couples, the alternation of seasons, the distinction of clans, and the categories of nature revolve. Just so, Durkheim had used rhythm as a means of anchoring the natural and the supernatural to their common origins in the social:

> For example, at the base of the category of time is the rhythm of social life; but if there is a rhythm in collective life, there must be another one in the life of the individual, and more generally, in the life of the universe. The social rhythm is simply more sharply and clearly delineated than the others. Similarly, we will see that the notion of

categories took shape on the basis of the notion of human groups. But if people form natural groups, there must be, among things, analogous yet different groupings. These natural groups of things are [what we know as] genus and species.[42]

One can scarcely imagine an example more obedient to its model than Granet's *Shijing* is to his analysis. The loss of individuality involved in our coming to terms with this "primitive" poetry is balanced by a gain: what is privation for the literary reader is a confirmation for the analyst of society. The impersonality of this poetry allows it to transmit an image of the social whole—what Marcel Mauss would later call a "total social fact"—an image that is barely distinguishable from nature (as understood by Chinese civilization).[43] Granet invites us to sacrifice the defining characteristics of poetry as we know it in modern societies so as to win access to the persuasive power of the collective, the consensus within which the stakes of "poetic jousts" are circumscribed. When "rhythm is all," poetry becomes a communal dance. Entry to this poetic commons is made possible for Granet only by dispossessing poetry's previous owner, the creative individual, hence the cascade of negations that forms Granet's appreciation. But that is not the only way to imagine the commons.

The Oral Style

Paulhan's surmise of "a language consisting of two or three hundred rhythmic phrases and four or five hundred verse-types, fixed once and for all and passed on without modification by oral tradition" received a further transformation in the idiosyncratic work of Marcel Jousse, S.J., *Studies in Linguistic Psychology: The Rhythmic and Mnemotechnic Oral Style of Verbo-Motor Individuals* (1925).[44] With Jousse, rhythm is not the skeleton of language stripped of its content, but the enabling medium of content. Jousse rewrites Paulhan:

> One might imagine a language consisting of two or three hundred rhythmic phrases and four or five hundred {Rhythmic Patterns}, fixed once and for all and passed on without modification by oral tradition. {Personal} invention would then consist of taking these {Rhythmic Patterns} as models and fashioning new {Rhythmic Patterns} in their image, {introducing propositional clichés as their Balancings}, {Rhythmic Patterns} having the same form, rhythm, structure, {the same number of words} and, so far as possible, the same meaning. Such a language would quite closely resemble the language of {the Oral Rhythmic Style in a still spontaneous Reciting milieu}: its {pat-

terns or Rhythmic Patterns} are proverbs, and its {Oral Compositions are} imagined in imitation of these proverbs, reproducing them in hundreds of new copies, stretching them out or shortening them, setting them amid other differently rhythmed phrases.[45]

Jousse's first book is mainly composed of citations from other authors, arranged and adjusted to compose the outlines of his ideas, as if Jousse were building an immense hain-teny that took the sciences of psychology, ethnography, physiology, philology, theology, and sociology as its proverbs. (Like every hain-teny, it has an antagonist, but of this more in due course.) Terms in the cited texts are changed in order to achieve a consistent terminology: thus Paulhan's "verses" become *schémas rythmiques* ("rhythmic patterns"), "poems" become "oral compositions," and so on. Other observations are inserted by Jousse in order to reinforce his thesis of the universal and compelling characteristics of memory and recitation in preliterate societies. Thus Paulhan, speaking through Jousse's revisions, emphasizes the rare "{personal} invention" of new verses in a traditional medium and affirms the place of hain-teny as an example of "{the Oral Rhythmic Style in a still spontaneous Reciting milieu}." And where Paulhan has omitted a detail that counts for Jousse's theoretical expansion, the latter does not hesitate to supply it: in a poetic joust the participants must "tak[e] these {Rhythmic Patterns} as models and fashion new {Rhythmic Patterns} in their image, {introducing propositional clichés as their Balancings}."

After commenting on Paulhan's hain-teny and a traveler's report on public speaking among the Bedouin, Jousse pursues:

> "Discourses of this type were compared to 'threaded pearls.' Such is, from one end to the other of {the *Quran*, or the Recitation}, the character of its sûras" (Letourneau, 272–273), and the same character is seen in the subtle Wisdom of Words among the Merinas; those Hebraic "enchainments" (*shir*: see Jastrow), whose names ring so strangely in translation: *Burden, Yoke, Comparison*, etc.; in the Counterweights of the Ethiopians, the Recitations of the Assyrio-Babylonians, the Egyptians, the Indians, the Chinese, the Australians, the Americans—these immense ethnic laboratories where the Psychology of Recitation, disengaged from the categories of any particular rhetorical tradition (no matter how developed), discovers with increasing interest the great universal laws of the Oral Style.[46]

Jousse enlarges considerably the poetic commons. Each culture has its vocabulary of a few hundred rhythmic phrases, its habits of linking and counterposing them, but all cultures are "laboratories" where one subject, "the great universal laws of the Oral Style," is explored. On the condition, however, that the laboratory remain a "still spontaneous Reciting milieu": for we who have acquired the habit of storing our memories in rolls, slabs, and leaves of external matter have allowed the pages of the great primordial Book to come unbound, to lose their order and coherence, a loss which Jousse seeks to repair by rediscovering the "laws of the Oral Style." The scope of the commons that concerns Jousse is panhuman.

> "The result of the studies so far prosecuted," says Mallery, "is that what is called *the* sign language of Indians is not, properly speaking, one language, but that it and the gesture-systems of deaf-mutes and of all peoples {and our own far atrophied mimicry} constitute together one language—the gesture speech of mankind—of which each system is a dialect."[47]

The oral style, for Jousse, is rooted in gesture and mimicry. All language is primarily gesture; both language and gesture demonstrate the properties of the human animal. (Thus Jousse's investigation finds its core in the two issues excluded from the debates of the Linguistic Society of Paris: the origin of language and the possibility of a universal language.)[48] At the base of the universal "mimic language" is the reflex arc, an automatic, physiological response to external stimulus that "exists in all living beings and is more fully realized in man to the degree that he is closer to the animal, that is, to the degree that the superior activity of the mind, which perturbs automatic operations, is neutralized."[49] Such is the case of peoples who have remained "spontaneous"—in another terminology, "primitives."[50] Such is also the case of the parts of the human organism that are below the threshold of consciousness: those under the control of the autonomous nervous system, for example.

> Organic life . . . rests on the indefinite play of organic reflex {gestures} responsive to internal stimuli. It is these that perform the ceaseless dynamism, {the perpetual gesticulation}, of the vital functions of assimilating and excreting food, the circulation of the blood, etc. All this is done by {gestures} separate from our will, but not separate from one another or from organic necessities . . . . Here we see the prime case of physiological automatism, most strikingly exemplified by the prodigious activity of the heart, so energetic, so tireless and so rhythmic.[51]

Intellectual life is organic life refined, reflected, and specialized. "In other terms, there are no states of consciousness that are purely observational: our observational states are accompanied by movements, {by gestures}, and thus by tendencies . . . Movements form the ground on which {and with which} consciousness embroiders its patterns."[52] Our body is "made of an infinity of automatic rhythms playing at different frequencies."[53] A new "Psychology of Gesture" is needed which will "take for its object the study of human behavior, {of gesticulation}, the study of partial {gestural} movements, and the general stances or motions of the whole body whereby an individual reacts to the effects that the various objects surrounding him have on him."[54] Any outside stimulus acts on the organism like a "detonator," a releaser of stored energy.

Memory, so often conventionally described in terms of "images," will be rephrased in Jousse's physiology of culture as a repertoire of relived actions. "'*Images do not exist*. This term must disappear from our language' (Moutier) . . . . 'For motor activity, {perpetual, microscopic or macroscopic gesticulation} permeates all of psychology' (Ribot)."[55] Bergson too had noticed that the role of repetition, in memorization, was to "employ ever more movements in the perpetuation of [a spontaneous recollection], to organize these movements, and, by fashioning a mechanism, to create a bodily habit" that would keep the memory at the ready.[56] Rhythm is precisely this recruitment and organization of movements in the service of memory. "It is certain, then, that the tendency to rhythm is a primary manifestation of the human brain, one with roots in organic life itself. Thus this tendency must be one of the last to disappear in cases of collapse or incomplete development of the mind {—that willful regulator and perturber of energetic explosions}. . . . Drunkards . . . insane people . . . degenerates . . . idiots . . . and the mentally retarded" attest to the survival of rhythmic behavior despite the wreckage of the conscious mind.[57]

The villagers observed by Jean Paulhan in Madagascar were not precisely poets—or degenerates for that matter—but adept performers in a memory-based civilization of the oral style. "'According to the method of the Orient . . . the child . . . repeats in cadence with his fellows {the recitation of rhythmic phrases} until he knows them by heart' (Renan). Naturally, 'this is the Orient, all right: these gesticulations, these cries, this rocking back and forth, these phrases obstinately repeated on the same monotonous rhythm' (Tharaud), but it is also, as we will see, the teaching method most deeply and universally fitted to the laws of the human organism."[58] Paulhan's observations were accurate as far as they went. He had noticed the traditional character of the verses chanted by the debaters in hain-teny, the family resemblance that made so many of the lines slight variants on other lines, their placement in a rhythm of call-and-response, their stereotyped content with a consequent metaphorical applicability to a wide variety of situations, and above all, their status as public property (so that even a newly invented line was quickly put into circulation and added to the communal stock). But Paulhan had not stated outright the purpose of the mechanism, as Jousse, in assimilating Paulhan's example to his own "laws," will do. The oral style with its rhythmic and mnemotechnic properties is a physiologically based media device evolved by people of various cultures in order to make information memorable and transmissible. Knowledge is a sum of learned gestures, including verbal gestures such as poems and proverbs. Education is the process, an outward, visible, physical process, whereby we internalize them.

Consider again Paulhan's hain-teny with their basis in "two or three hundred rhythmical phrases." Such "typical rhythmic patterns," in Jousse's view, are "an automatic trigger, available to free will but sparing of its energy," installed in the mind of every competent speaker; they are a store of "semiological gestures laid out in readiness."[59] Through repetition and rhythm, the teacher (bard, rabbi, prophet, or tribal historian) "plants, by persuasion or by force, his propositional Balancings into all the young muscles [of the pupil] . . . The first rhythmic Pattern was a rhythmic-didactic Pattern."[60] Thus "force" or trauma, it is suggested, may be mnemonic.[61] The pupil, properly taught, is like a wound-up clock whose movements can be precipitated (*déclenchés*) by the slightest invocation of the learned patterns. The best rhythmic-didactic pattern is the one that takes over the reciter's body most efficiently, brings it to the greatest degree of "automatism."[62] Corresponding to the discovery of a collective poetic reservoir in Paulhan is Jousse's desire to give the greatest possible powers to memory, to the preinscription of the traditional reflexes, to automatism, and to relegate consciousness to as slender a margin of activity as possible. The perfect type of knowledge in Jousse is reflex action, a latent response that awaits only a stimulus to make the texts "perform themselves" in their performers, and link one "rhythmic pattern" to the next by echo and parallelism. "The triggering of one pendulum-swing by another, the mutual call of two or three Balancings memorized together in the past in the same rhythmic Pattern, is so automatic that often the oral composer has only to let the automatism run by itself."[63] As one embittered colonist put it: "Should we call [Nigerians] born orators, masters of language? It is far more the case that language masters them and plays with their naïve intelligence."[64] But to Jousse it was no shortcoming for a Reciter to be mastered by the language of the Recital.

Marcel Jousse is perhaps the most impassioned discoverer of oral tradition. It was through his American auditor Milman Parry that a carefully filtered subset of his ideas reached wider currency. But we must be careful not to assume that "oral literature" was simply a fact waiting to be discovered by Jousse, Parry, or another: it had to be elaborated in the course of a history consisting of choices, exclusions, chance events, outside pressures, and the like. Orality for Jousse is not merely a fact about the origin of texts: This origin gives its texts a particular kind of organization and predisposes them to a particular kind of meaning. If formal analysis can uncover these particularities, then whether a text is "oral" should no longer be merely a circumstance observed by the text-collector. The distinctiveness of orality is the main assumption Jousse and Parry hold in common. Moreover, Jousse speaks of the oral style as a domain of "universal laws"; though Parry and his lineage are less grandiloquent, they likewise do not hesitate to diagnose "features of oral poetry" in places and times far removed from one another (Homeric Greece, the interwar Balkans, Anglo-Saxon England, and so forth).

But what are the outlines of this domain? How is it to be distinguished from other neighboring textual domains? Do its laws apply exclusively to oral texts (assuming we know how to identify them)? What laws apply in other correlated domains?

At times, Jousse answers these questions with the by now familiar denunciation of writing as the agent of denaturalization, the artifice that kills living memory. In this perspective, the oral style is the antithesis of "papyvorous" habits of thought and expression. The essentialism and primitivism, not to say logocentrism, of this move are patent.[65] But Jousse's antipathy to writing does not preclude his speaking of the oral teacher as "inscribing" his message on the lips of the learner, or of orally transmitted folk songs as the "living press" of their times.[66] The established phrases and sayings of an oral tradition are known as "clichés," a word Jousse takes from the history of typography: a block of text or image designed to be reproduced as a unit, a "stereotype," which word originally designated a metal casting made from a page of already set-up type.[67] As the metaphors tacitly concede, orality, writing, printing and the like have one great fact in common: the characteristic of being media. And many of the texts to which Jousse seeks to attach the name of oral style have long circulated in manuscript and print, whatever their earliest origins: Homer, the Bible, the Buddha's sermons. So a rejection of writing will not by itself circumscribe the oral field. To seek the outlines of the domain of the oral style, we would do better to discover them empirically, by trial and error and misunderstanding—by allowing ourselves to be perturbed.

Thus Jousse transcribes a number of travel reports in which a European visitor asks a question of an individual trained in an oral culture and gets a singsong recital in return. The questioner expects an answer that will be relevant to the immediate circumstances and uttered by an individual in context, and hears the voice of collective memory instead. This "infuriates" (*agace*) the literate Europeans who think they are no longer talking to a conscious individual as soon as the interlocutor's speech takes on the contours of an insistent, relentless, patterned litany.

> "An aged Turk squats on his mat at the foot of the column on which the sacred trophies are suspended. He is praying. He mutters again and again, in a stuttering tone, wild-eyed, swinging his head and torso back and forth, the believer's creed: '*La ilaha il Allah!* God alone is great!' This monotonous muttering, filling the whole mosque, is annoying (*agaçant*) and disturbs Brother Liévin's historical presentation. A friendly imam tells the good Turk to moderate his fervor. But he does not hear a word of it and continues his hoarse litany: *La ilaha il Allah!* The imam shouts more loudly and seizes him to put a stop to it, but to no avail. The priest becomes stubborn and gives the old fellow a push: the machine keeps going, without a sign or a glance or a facial expression to suggest that the old man has the slightest concern for what we are trying to tell him" (Monseigneur Landrieux). It is such contacts with the life of ethnic groups other than our own that build up in us new mental attitudes.[68]

We have here the clash of two sociologies of oral expression. Literate people misunderstand the workings of this impersonal, self-perpetuating memory that resides in (what seem to be) persons. "The machine keeps going." Their irritation is the sign that their linguistic economy has been perturbed. Perhaps only an optimist can expect "new mental attitudes" to arise from the conflict.

"This is the Orient, all right"; "The machine keeps going." Formulaic style dehumanizes; or (on another reading) it furnishes a pretext for dehumanizing speakers in the "oral style," when observed by literate speakers. Here is a hint of the broad division of humanity into "oral" and "written" cultures, laid out in tabular fashion by Walter Ong; but the difference in style of presentation matters. The antithesis here arises out of a dramatic situation, one in which formulaic utterance frustrates and creates a scandal for the literate visitor who can no longer experience it sympathetically, but sees in it only loss of selfhood and meaningless repetition. (Granet, reading the "stereotyped epithets" of early Chinese love poetry, expressed precisely this *horror vacui*.) Insofar as people are engaged with different media regimes, they will have different relations to language, different modes of consciousness, that do not mutually translate, but cause blockage, frustration, and scandal. Yet the scandal can become a discovery technique. Jousse's transcribed anecdotes show the mutual constitution of "orality" and "literacy": Neither exists except as resistance to the other.

If the discourse of "verbo-motor" subjects seems like madness or machinery to the literate outsider, this is for the good reason that they are not speaking on their own behalf, but transmitting dictation. "'It follows that {as we have seen} the {oral composer} thinks and remembers with his {throat and mouth} muscles' (Arréat), with the gestural clichés that his social milieu has wound up in him. We should then expect to hear identical traditional formulae pronounced by all members of a given linguistic milieu, and by their descendants."[69] Dictation, that is, from the past: for Jousse insists that these traditional formulas are designed to ensure their own survival in the risky milieu of human memory. They are meant to stay the same over time and to reproduce themselves faithfully in different individuals.

This mode of preservation and delivery accounted for

"the {Recitations} of the [South-Slavic] guslars, [which,] {analogous to the Recitations of Homer, of the prophets and the rabbis, to the Epistles of Baruch and of Saints Peter and Paul, to the delicate paired verses of the Chinese, and so on} are a juxtaposition of clichés, the number of which is relatively small. The expansion of any of these clichés is automatic, guided by fixed rules. Only their order may change. A good guslar is one who plays his clichés as we do cards, arranging them differently in function of the result he wants to achieve with them."[70]

In a milieu of the Oral Style, plagiarism and composition by clichés are nothing more than a greater or lesser density of usage of propositional gestures that are stereotyped and available for anyone to use. A good Rhythmer in the Oral Style is "one who plays his clichés as we do cards."[71]

Jousse, like Paulhan, thinks of tradition not as a long scroll to be memorized from beginning to end, but as a vast database or collection of short "bits" or "formulas," the spare parts of potential or realized traditional utterances.

"Despite the disapproval that the doctrine of the young Rabbi of Nazareth encountered," "{his} expressions are easily recognized as stereotyped formulas common to all Doctors, both Jesus and the Rabbis" (Buzy). . . . And the scholar who said that "the Sermon on the Mount was current in the streets of Jerusalem long before it was pronounced" by Jesus (Munk, cited by Rodrigues) offered this observation about the wandering clichés of a particular oral literature, just as he could have made it about the wandering clichés of all oral literatures, transcribed or not, by saying (*mutatis mutandis*) that the inspired Recitations of Homer or the mighty historical "chain" forged by this or that Afghan Rhythmer could be found on the lips of their contemporaries and predecessors long before they were ever spoken by the oral Composers themselves. Nothing {obviously} could be easier than "to recreate the Sermon on the Mount with materials of a previous era" (Munk, cited by Rodrigues). . . . [They were] mosaics made of oral formulas transmitted from one generation to the next.[72]

A gulf appears between language and formulaic language: once we are conscious of the latter, the Sermon on the Mount no longer "belongs" to Jesus. He merely activated it. In the same way, Paulhan's friend did not really say the word "ox." In traditional speech, words are not free, but bound in formulaic expressions; the formulaic expressions too are not free, but bound in sequences or compositions; and these compositions, like the formulas, are common property and cannot always be analyzed into meaningful components. The discovery of the commons operates a shift in discourse that reassigns most of our markers of meaning. It is not enough to ask, what did the author intend? One must also ask, what does the medium intend? The discovery of collective, generative, iterative textuality—in this instance, of orality—perturbs reading in specific and regular ways.

It is one thing to experience perturbation, another to account for it, yet another to seek to provoke it. The perturbation that Jousse, with his oral style, wanted to accomplish in the world of letters will be discussed in another chapter. We now move to another scene of the discovery of nonliterate letters, Milman Parry's journey from chaos to order.

Formula as System

The eminent comparative philologist Antoine Meillet let fall at least one scandalous statement in an otherwise exemplary scholarly career.

The Homeric epics consist from beginning to end of formulas trans-mitted by one bard to another. Take a passage at random: one quickly recognizes that it is made of verses or verse fragments that occur textually in one or more other passages. Even the verses that are not repeated, in whole or in part, have a formulaic character, and it must be a matter of mere chance that they are not preserved elsewhere. It is true that the line ἀλλὰ μάλ᾽ εὔκηλος τὰ φράζεαι, ἅσσ᾽ ἐθέλῃσθα *alla mal' eukêlos ta phrazeai, hass' ethelēistha* (*Iliad*, 1:554) recurs nowhere else in the *Iliad* or *Odyssey*, but the reason is simply that the bards had no other occasion to use it.[73]

The English philologist Arthur Platt could not believe what he was reading: "Things are said about the epic on page 61" of Meillet's work, he wrote, "which make one stare."[74] But Meillet's usual style was careful and modest, not provocative; what could explain this uncharacteristic moment of conjecture?

In a word, generativity. Reviewing Jean Paulhan's study of hain-teny in 1913 for the *Bulletin of the Linguistic Society of Paris*, Meillet had conceded that the book was "not exactly a work of linguistics. But it should be called to the attention of linguists because they will find there a remarkable example of a type of unwritten literature that must have existed in many places other than Madagascar, and which certainly contributed to the preservation of many archaic words and fixed turns of phrase."[75] Writing for an audience of historical linguists, Meillet emphasized the fossilized quality of formulae, the fact that they come ready-made to the contexts where they are used and are rarely if ever reformulated. They were inscribed, so to speak, on the memories of the former poets at some time in the distant past and have been faithfully repeated, in new combinations to be sure, ever since; hence their value for the antiquarian.

Ten years later Meillet took another step. To the long-standing contention that Homer's poetry had been an oral tradition fixed in writing late in its development, Meillet added Paulhan's new contribution, the idea of a reservoir of ready-made poetic phrases, accessible to the memories of all poets in the tradition, from which new lines and episodes could ceaselessly be built up, thus offering a new and provocative vision of ancient epic. The combinatorial view leads us to see the *Iliad* and *Odyssey* not as isolated masterpieces but as sample utterances with a common language between them, two outputs of an infinitely variable virtual system. Platt's incomprehension betokened the differing protocols of philology and linguistics as of 1923: the first committed to the interpretation of individual documents, the second already beginning to take a structural turn. Whose business was it to speculate about unwritten *Iliads* and *Odysseys*? Not the philologist's; perhaps not yet the linguist's.

While Meillet's provocative statement about Homer was still in press, he received a visit from Marcel Jousse, a Jesuit priest and ex-artillery captain who had been studying experimental phonetics with Pierre Jean Rousselot, psychology with Pierre Janet, and sociology with Marcel Mauss. In Jousse's recollection,

I can still see him leaning on the mantelpiece—I laid out for [Meillet] the laws of the Formula, or of Formulism . . . I said to him: "The proverbs of the Sarthe region are formulaic, the style of the *Kalevala* is formulaic, and there are many other examples, for example the style of the Hain-tenys of the Merina people . . . and the style of Homer has to be formulaic because it's an oral style." I rattled off my little piece and Professor Meillet said: "This is astonishing. Would you by any means have had access to the proofs I am just now correcting of my *Origins of Greek Meter*?" I said, "Professor, I would certainly have much to gain from reading them but I have not seen them yet."[76]

As if to fill out the suggestion in Meillet's 1913 review that the mechanism of formulaic composition "must have existed in many places other than Madagascar," Jousse cites the parallel cases of his home region in France and the Finnish national epic and seemingly predicts the conclusions of Meillet's work in press. This episode of mutual recognition pleased Jousse so much that he retells it at least half a dozen times in his recorded lectures on oral and gestural culture.[77]

When the young American classicist Milman Parry, bearing an M.A. from Berkeley, came to Paris in 1924, he intended to study with Victor Bérard, the translator of Homer, but found more common ground with the literary historian Aimé Puech and the historical and general linguist Meillet.[78] Parry's is of course the name most readily associated with the idea of oral poetry in the English-speaking world. Through Meillet, Parry would have learned about Jousse's synthesis of existing research on oral memory and recitation, published in 1925.[79] And at the conclusion of Parry's thesis defense in 1928, Meillet introduced him to Mathias Murko, a Slovenian scholar of folklore who had come to Paris to lecture on the epic poems recited by Yugoslav bards (a tradition repeatedly mentioned by Jousse).[80] If Parry's main contributions to scholarship are the theory of epic formula and his collection of contemporary South Slavic oral-traditional texts, neither was unprecedented, though Parry gave a new form to both.[81]

Parry's cast of mind was inductive, not theoretical. He brought a deep knowledge of the Homeric poems and matchless patience to an ancient problem—the problem of epic "style." Homer's use of language is "eminently rapid . . . eminently plain and direct . . . [and] eminently noble,"[82] but also at times redundant and inexact. He does not mind calling Aegisthus "blameless" in the same breath with which he goes on to tell how Aegisthus seduced Clytemnestra, a far from blameless act; his ships are "swift" whether they are in motion, beached, or at anchor; he may mention a "brilliant" moon that he has just described as invisible or qualify a riverbank laden with corpses as "lovely."[83] And Homer is profligate with his epithets beyond all requirements of sense, reminding us again and again that Achilles is swift-footed or Odysseus wily, whether or not there is anything in the immediate context that makes swift-footedness or wiliness relevant. Indeed, hundreds of verses and dozens of scenes are repeated word for word elsewhere in the poems. Henri Estienne supposed that Homer might have been parodying himself. Houdar de la Motte thought that the only explanation for the poet's "cold and tiresome epithets" was that he "enjoyed fattening up his opus with material that cost him nothing to produce, and that the pleasure of writing his own lines over and over made him unable to see how useless and purposeless it was."[84]

Commentators over the centuries, loath to admit that Homer might have talked nonsense, have invented strained and ingenious justifications for each of his irrelevant descriptions.[85] Perhaps it was ironic to call Aegisthus "blameless"; maybe the beached ships are "swift" potentially if not actually; maybe the river is supposed to be "lovely" and the inopportune word laments its disfigured state. When such ad hoc arguments fail to convince, the practice of readers through the centuries has been to disregard the epithets and stock narrations, to take them as mere idle decoration, *epitheta ornantia*, the customary grand-sounding verbiage of epic style, and nothing more. The "swift ship" is the epic ship; the "wine-dark sea" is the epic sea, and so forth. It does not do to ask Homer to make sense all the time. The tactful reader learns to stop asking such questions.

If the need is felt of a consistent explanation of the inconsistencies of Homeric poetry, the lazy reader is the very pattern of consistency. From long experience of "cases where no relation between the idea of the epithet and that of the sentence is possible," the typical reader is educated into an "insensibility" or "indifference" to the repeated adjective wherever it appears.[86] "Illogical usage is of interest here," says Parry,

> because it attests not only the ornamental meaning of the epithet, but also the poet's inattention to which name the epithet was to accompany. If the poet paid so little attention to the signification of δῖα [godlike] when he used it for Clytemnestra, it is obvious that he was paying it no more attention when he used it for Odysseus or for Priam or for Alexander, or even for that Aretaon who appears in the *Iliad* only to be killed by Teucer. . . . His audience found nothing out-of-the-way or ludicrous in [such illogical usages] because it never occurred to them for an instant to analyse the noun-epithet formula.[87]

"The epithet has no bearing on the idea of the sentence."[88] "Homer's listeners demanded epithets and paid them no attention."[89] They heard Homer in a state of distraction, as if their minds were elsewhere.

From the point of view of its meaning, then, Homeric language when looked at closely is a loose and garbled thing. But set against its semantic vagueness is a great rhythmic precision. Parry discovers in Homeric language not sense, but order. Take two recurring lines from the *Odyssey*:

Ὣς ὁ μὲν ἔνθα καθεῦδε πολύτλας δῖος Ὀδυσσεύς (e.g., *Od.* 6.1)
Hōs ho men entha katheude polytlas dios Odysseus
(so there lay down the much-enduring and godlike Odysseus)

and

Ἂν δ'ἄρα διογενὴς ὦρτο πτολίπορθος Ὀδυσσεύς (e.g., *Od.* 8.3)
An d'ara diogenēs ōrto ptoliporthos Odysseus
(now rose up the divinely born Odysseus sacker of cities).

On looking more broadly, we find that Odysseus always goes to sleep described as "much-enduring," *polytlas*, and "godlike," *dios*, but he awakens as a "sacker of cities," *ptoliporthos*. Does the poet mean that a good night's sleep restores a tired man to the condition of an invincible warrior? Or that Odysseus is somehow special in this regard? To pursue the meaning of these phrases too far might lead us to look for the difference between them in the wrong places; to pursue it in analogous passages (sleep episodes, classified by hero and by result) might lead us to conclude that there is no difference *in meaning*. A wider survey of the two epics shows that

> If we take all the noun-epithet formulae for Achilles, in all five grammatical cases, we shall have 45 different formulae of which not a single one is of the same metrical value in the same case as any other. If we take all those which are used for Odysseus, we shall find 46 different ones, and of these only 2 are of equivalent metrical value. . . . Like systems can be established for *horses*, for *the human race*, for *the Achaeans*, for *ship*, etc.[90]

Meaning is too crude a tool for drawing the distinctions and correlations that need to be made among epic modifiers. The difference between *ptoliporthos* and *polytlas dios* is not the difference between the meanings "city-sacker" and "much-enduring godlike one," but the difference between their metrical shapes, and, as a consequence, the difference between two sorts of place that they can fill in a dactylic hexameter (˘ ˘ ¯ ¯ ˘) and ˘ ¯ ¯ ¯ ˘). Or take the pair πτολίπορθος and πολυμήτις ("city-sacker" and "wily"), both appropriate epithets for Odysseus under any circumstances: their relevant difference is not a matter of metrical feet (they have the same shape) but that between an initial double consonant, which lengthens a short vowel preceding it, and the single consonant.[91] If you wish to know why the Odysseus who rises up (*ōrto*) in the morning is a sacker of cities, the answer is: because the verb "he rose" ends with a short vowel where a metrically long syllable is required, and the *pt–* of *ptoliporthos* makes it long, which the *p–* of *polymētis* cannot do. Defining the formula as "an expression regularly used, under the same metrical conditions, to express an essential idea,"[92] Parry described epic language as "a system of formulae which is a set of subsystems"[93]—the "Achilles" subsystem, the "ship" subsystem, the "spoke angrily" subsystem, and so forth. The set logically precedes its members: "Often we have the proof for a set of formulae without having it for each element."[94] Once the right analytical filter is applied, the Homeric formula-system exhibits hundreds of correlated units arrayed along two complementary dimensions of "extension" (the number of available alternatives for a single basic meaning) and "economy" (the absence of metrically redundant epithets for that same semantic slot).[95] When the elements "cannot be replaced by any other form," which is not always but surprisingly often the case, "we are in the presence of a rigorously fixed system showing both great extension and complete absence of any superfluous element, indications that the whole system should be considered traditional."[96] Extension and economy together show that the epithets are, *qua* elements of a system, differential entities, as Saussure would have said: within the group established by their loose semantic affiliation, the "most precise characteristic" of each "is in being what the others are not."[97]

The reader who would seek Homer's meaning finds only a mass of redundancies, approximations, and near-synonyms; Parry, the reader who looks for structure, replaces this nebula with a system of differences in which every noun-epithet formula has its unique architectural value. Word choice matters, but the meaning of the words is not all that matters in word choice. Or to put it another way, Parry admitted that Homer often does not make sense, but he also showed that sense is not primarily what Homer was making. The rhythmic properties of the words and phrases are a means of achieving order, and to ask for too much sense is to upset the balance of order. Ordinary language (say, the language of a translation) veils this order. Like the proverbs initially misunderstood by Paulhan as being statements composed of separable, individually referring words, the formulas take ordinary language as their raw material and from it build up another code with its own organizing principles, sus-pending in the process certain aims and properties of ordinary language. As outsiders, as nonpractitioners of the "sacred language," we are easily misled and frustrated by what in it seems to us mechanical, irrelevant, alienated. A new aesthetic suited to these new building blocks will resolve this tension.[98]

Parry entered Homeric studies as a literary scholar and emerged as an unanticipated kind of linguist. Indeed, between Parry's first and second articles in English, the phrase "Homeric style" was replaced by "the Homeric language."[99] The change consists in this: The traditional distinction between "language," the common property of all speakers, and "style," as individual endowment, disappears, to the benefit of "language" or now "a poetic language." "The fixed epithet did not so much adorn a single line or even a single poem, as it did the entirety of heroic song."[100] In far greater detail than any of his predecessors, Parry carries out the project of reori-enting the study of poetry from the poet to the tradition, from the individ-ual document to the shared code, from the attested corpus to the virtual program that must have generated it from a latent stock of material.

Thus, although the formula was already formulated at the time of Parry's arrival in Paris, through him it acquired internal consistency: a functional differentiation into parts and a set of organizing laws dis-tinct from those presiding over the meaning and grammar of ordinary language. "The essential feature of such poetry," he later argued, "is its oral form, and not such cultural likenesses as have been called 'popular,' 'primitive,' 'natural' or 'heroic.'"[101] This advanced the work of reconceiving orality to which Paulhan had given a singular impulse. The tacit defini-tion of orality as the absence of writing or the failure to achieve the sta-bility associated with writing began to be replaced by a set of substantive characteristics: formularity, generativity, collectivity. Oral composition acquired a structure of its own—a structure different from those of either ordinary language or written composition, as scandals of misunderstand-ing in the contact zone between orality and literacy showed. Parry gave a finer grain to that structure, but as if by compensation, deepened the gulf between orality and writing.

Langue, Parole, and Constraint

What Paulhan put in motion and Parry pursued in meticulous detail was a formal, not circumstantial, definition of oral composition. When certain texts came under suspicion of having been composed of interchangeable parts rather than springing from a single creative mind, the interchangeability of the parts suggested that the text handed down by history was only one of their possible combinations. Such texts stood at the brink of a vast and unsettling redefinition, of which Marcel Jousse gave the most adventurous example. In 1929 two Russians, a folklorist and a linguist, both working in Prague, gave a new name to the virtual, generative dimension that Paulhan, Jousse, Parry, and a few others had detected in the background of their chosen texts. They wrote that

> one of the essential characteristics differentiating folklore from litera-ture is the concept of the being of a work of art [*der Begriff des Seins eines Kunstwerkes*].
> In folklore the relationship between the work of art and its realiza-tion, i.e., the so-called variants of the work in the performances of different persons, is completely analogous to the relationship between *langue* and *parole*. Like *langue*, the work of folklore is extra-personal and has only a potential existence, it is only a complex of particu-lar norms and impulses, an outline of the current tradition that the executants animate with the embellishments of individual creativity, just as the creators of *parole* do in regard to *langue*.[102]

A myth, a ballad, or a folktale may never be recited twice in exactly the same way. Collectors of such material are aware of variation and deal with it variously in their printed editions, usually by choosing one version as the standard, by recording alternative wordings in their footnotes, or by arranging a series of versions as testimony to temporal or geographi-cal diffusion. But what Bogatyrev and Jakobson were proposing must be imagined differently from a table of variants. If two, or fifteen, or a hun-dred reciters individually "alter" a text, but we still agree that they are performing the "same" text, then the criterion for identity cannot be the sameness of wording; it must be something more abstract, a shared out-line or recipe.

The problem of identifying the authorship and identity of an oral text was familiar. George Lyman Kittredge, introducing a selection of ballads in 1904, had offered,

> Here the mere act of composition (which is quite as likely to be oral as written) is not the conclusion of the matter; it is rather the begin-ning. The product as it comes from the author is handed over to the folk for oral transmission, and thus passes out of his control. If it is accepted by those for whom it is intended, it ceases to be the property of the author; it becomes the possession of the folk, and a new pro-

cess begins, that of oral tradition. . . . We may compare the processes
of language. A word is created by somebody. It then becomes the
property of the whole body of those who speak the language, and is
subjected to continuous modification. . . . It follows that a genuinely
popular ballad can have no fixed and final form, no sole authentic
version. There are *texts*, but there is no *text*.[103]

Saussure, the inspirer of the movement that would be named "structuralism" only in 1929, had framed the issue of identity and difference
ingeniously in the *Course* published by his students in 1916.[104] Admitting
that no two renditions of a sound or a word are ever absolutely identical,
Saussure held that the analysis of language could not hope for anything
better than to discover enduring *functions* which a variety of empirical realizations might occupy.[105] The 8:15 Paris to Geneva express, in Saussure's
famous comparison, might be composed of different cars and personnel
every day, might never depart at precisely 8:15, but what makes it that
express and no other is the mustering of elements to accomplish a certain
task (as opposed to the task known as the 7:39 local, and so on), not the elements themselves. Variability was everywhere in language, especially as
historical linguists and dialectologists saw it, but the description of sound
systems, meaning systems, and grammatical forms in terms of function
resolved those differences and proposed a relatively stable set of objects
for analysis. The name Saussure gave to the interconnected set of functions underpinning the speech of a given community was *langue*.

Jakobson and Bogatyrev do not seem to have known about Milman
Parry in 1929, but their Saussurian analogy economically restates the purposes of Parry's early research.[106] The point of Parry's two theses for the
Sorbonne doctorate, seen in Jakobson and Bogatyrev's terms, was to reori-
ent the study of Homeric epic from the *parole* of its two main documents
to its *langue*.[107] The mode of analysis is no longer hermeneutic, but gram-
matical; not primarily concerned with meaning, but with structure. Many
meanings (or failures of meaning) are there explained as a consequence
of structure, of the *langue* held in common by all epic composers rather
than a specific bard's artistic intention. Like the Saussure of the *Course
in General Linguistics*, Parry sharply differentiates between the tasks of
synchronic functional description and diachronic historical explanation,
giving logical priority to the former.[108] It is after all only thanks to the
pressure of the system (a panchronic constant) that an element of diction
finds or fails to find its place; the events of its introduction or elimina-
tion are for Parry historical accidents.[109] If the history and development
of the formulaic system must be explained, Parry reaches, as if in echo of

Saussure's approach to similar problems in lexical history, for the workings of analogy—that motor of linguistic change which is nonetheless
"wholly grammatical and synchronic."[110] Finally, the motif of the inability
of meaning to reveal order in language recalls Saussure's reworking of the
neogrammarian doctrine of the "blind" and "exceptionless" character of
phonetic shifts.[111] Differences in form—phonetic form for the neogrammarians, metrical shape for Parry—explain differences in meaning, but
not the other way round.[112] Reminiscent of Saussure's commandment to
his students—"The beginner must compose his own grammar on the basis of a determined text, imposing on himself a law never to depart from
it"[113]—*L'épithète traditionnelle* is a long demonstration that the "grammar"
of Homer cannot be confused with the grammar of Greek outside of epic,
despite the constant contact between the two that allows epic to address a
public largely composed of non-bards.

In reaction to a long-standing habit in Slavic lands of elevating folklore
to the status of literature, Bogatyrev and Jakobson insist on the differences between folklore and written texts, to the point of drawing a neat
antithesis: "From the folklore-performer's standpoint, the work is a fact of
langue, i.e., an extra-personal, given fact, independent of the performer,
even if the fact allows for deformation and the introduction of new poetic and occasional material . . . An essential difference between folklore
and literature is that folklore is set specifically toward *langue*, while literature is set toward *parole*."[114] This must have been confusing for their
earliest readers. For oral literature would seem to be entirely committed
to *parole*, to the event of speech, that fugitive articulation of the common linguistic heritage in the mouth of one individual.[115] Writing, by
contrast, stores the moment up and makes it available for later circulation. But this likely misunderstanding would entail missing the structural
specificity of folk or oral texts: their repetitiveness, their recycling of past
models of the text, their scant individuality. The performer is bringing a
potential work out of the invisible storehouse of tradition—*langue*—and
will do so again, somewhat differently, at the next opportunity. It is the
writer, according to Bogatyrev and Jakobson, who operates in the realm
of *parole*, crafting his personal utterance once and for all and, as a condition of success, differentiating it from the other utterances in the literary
field. As a result, literary history can be seen (as in Russian Formalist accounts) to evolve, to renew itself through repeated offenses to good taste
and precedent, whereas oral literature is imagined as a ceaseless return to
the norm.[116]

Applying the analogy of *langue/parole*—the communal system of language contrasted with the individual's occasional uses of it—might sug-

gest a new task for the folklorist: constructing the rules or the recipe that different performances of a folkloric text have in common.[117] This is not a task that the literary historian or critic need be concerned with, though it is a fair description of the efforts of Paulhan, Jousse, and Parry. But Bogatyrev and Jakobson seek the *langue* elsewhere than in tabulations and correlations. If the work, in terms of its ontological status (*der Begriff des Seins [des] Kunstwerkes*), is not to be confused with any of the performances of that work, where does the work reside? In a culture of inscription and archives, the identity of "the work" can indeed be located in some durable artifact: in a musical score, in the author's manuscript, in the first publication or recording, in a copyright application. But a work that circulates by word of mouth leaves no such durable sign of its beginnings, and there is no way to track a series of performances back to their original. For this conundrum, Bogatyrev and Jakobson propose a sociological solution. Between performances, the work exists as a relation between performers and audience, as reception and expectation.

Consider, for the sake of contrast, the case of a long-ago poet whose verses, forgotten on the shelves of a few libraries, are one day rediscovered and become the inspiration for a new style; or that of a memoirist whose secrets from behind the scenes of palaces and parliaments were too scandalous to divulge at the time of their writing, but furnish later historians essential material for understanding the age. For a longer or shorter period, those texts had no reception at all. Such gaps, routine in a literary and archival culture, are impossible for a folklore milieu where the means of preservation is nothing other than active memory, stirred by frequent revival in performance. Can we imagine the Comte de Lautréamont in an oral culture? "Upon his death his works would have disappeared without a trace." For "the existence of a folkloristic construct as such does not begin until it has been accepted by a certain community,"[118] and there might have been no community in 1870 ready to receive and preserve what would have been considered his monstrous, blasphemous ravings. An oral Lautréamont would have been ignored as a madman, forgotten, censored, and never would have had the chance to be rediscovered by the 1920s Surrealists.

Thus the fact of orality prompts a different way of recounting cultural history, one in which the libraries are always on fire and survival is the exception. Formulaic language and adherence to convention serve the purposes of memorization and protect an oral text, by anticipation, from the censorship that awaits all incorrect recitals. Poetic texts formed in a literate culture are constrained by conventions such as rhyme and meter, as are oral texts. But according to Jakobson and Bogatyrev,

It would be ambiguous to speak of identical forms in folklore and literature. For example, meter, a concept which at first glance seems to have the same meaning in both, is in reality profoundly different in terms of function. Marcel Jousse, the sensitive [*feinfühlige*] scholar of the oral, rhythmic style (*style oral rythmique*), considers this difference so significant that he reserves the concepts 'meter' and 'poetry' for literature alone. . . . Jousse masterfully uncovers the mnemotechnical function of such rhythmic schemata. He interprets the oral rhythmic style in a "milieu de récitateurs encore spontanés" [milieu of still spontaneous reciters] in the following manner: "Let us imagine a language with two to three hundred rhymed sentences, with four to five hundred types of rhythmic patterns, that are precisely fixed and transmitted without the modifications of oral tradition: personal invention would then consist in using these rhythmical patterns as models by analogy and with the help of sentence-clichés shaping other rhythmical patterns of similar form, with the same rhythm, the same structure . . . and so far as possible with the same content." The relationship between tradition and improvisation, between *langue* and *parole* in oral poetry, is clearly delineated here. In folklore, meter, stanzas, and even more complicated compositional structures are both a powerful support of tradition and an effective means of improvisational technique (the two, of course, being closely related).[119]

In the translation from French to German, and also in the translation from Paulhan's ethnographic poetics to that of incipient structuralism, a syntactic ambiguity has triggered a misreading: *Schèmes . . . fixés pour toujours, transmis sans modification par la tradition orale* means "patterns . . . permanently fixed, transmitted by oral tradition without modification," but *Sätze . . . die genau fixiert, ohne die Modifikationen der mündlichen Tradition überliefert sind* means "sentences . . . that are precisely fixed and transmitted without the modifications [characteristic] of oral tradition."[120] In the French original, oral tradition transmits the two or three hundred phrases without modification (and that lack of modification is notable, contrasting with "personal invention"); in the German version, "modifications" are what an "oral tradition" always produces, and the inherited corpus of two or three hundred phrases must stand somehow outside orality, that stream of mutability, in order to be "precisely fixed" and transmitted without a slip, inevitably raising the question: How?

Paulhan and Jousse, we see through the slight difference imparted by the German translation, were working from an assumption, indeed the proud assertion, that societies of oral tradition are static and tend to maintain what has been passed down. Bogatyrev and Jakobson start from the opposite hypothesis, that the medium of orality is defenseless against change and must enlist other forces to guarantee the preservation of cultural information. The task of ensuring stability falls to the social frame.

The career of an oral text—shall we adduce the *Iliad* or the Sermon on the Mount?—"does not begin until it has been accepted by a certain community," Bogatyrev and Jakobson reason, "and the only aspects of a construct that persist are those that have been appropriated by this community."[121] A folk song may acquire variations—"improvisations"—over the course of its history, but not all of those variations are equally apt to survive. Innovations must be true to type, so to speak, with the definition of the "type" left up to the judgment of the community, passively and unconsciously exercised through the forgetting of malaprop additions. "Everything rejected by the milieu simply does not exist as a fact of folklore. It falls into disuse and dies out."[122] The model of the work as a productive, generative *langue* stands in a constant relation to its negative image: forgetting. Bogatyrev and Jakobson call this "the preventive censorship of the community" (*die Präventivzensur der Gemeinschaft*) and speak of "the inseparable fusion of censorship and work characteristic of folklore." Their terms are unusually peremptory: while "a literary work is not predetermined by censorship and is not completely determined by it," in folklore, "censorship is imperative and is an indispensable prerequisite for the genesis of works of art."[123] If the *langue* of a folklore production is imagined as including a certain set of formal and thematic features that suffice to define the piece in its successive performances (its instances of *parole*), "censorship" symmetrically excludes what is inconsistent with those features; if a particular instance of *parole* should conflict with the norms set by *langue*, it will be treated as a mistake and ignored. By such means cultural objects are maintained over time without the use of writing. "The absolute dominance [*absolute Herrschaft*] of preventive censorship, rendering fruitless any conflict that the work may have with that censorship, creates a special type of participant in poetic creation and forces the individual to abandon any attack aimed at overcoming censorship."[124]

Having reconceived literature and oral tradition in terms of their distinctive mnemonic economies, Bogatyrev and Jakobson then move to show the power of their model to rebut and reframe certain long-debated terms in folklore studies. "The Romantic theorists . . . were correct in emphasizing the communal character of oral poetic creativity and in comparing this with language."[125] But Bogatyrev and Jakobson balanced this assertion (often paraphrased in the saying, "*das Volk dichtet*," "poetry comes from the People") with another claim useful for extricating the concept of folk creativity from populist mystification. Cases of "*gesunkenes Kulturgut*" (devaluated cultural goods)—a shorthand for the idea that folk culture consisted of fragments of élite, literate culture that had been acquired and reshaped by the lower classes—demonstrated that what was specific to folklore was not a matter of origin, but the way that it reshaped its materials in circulation by the processes of reproduction and censorship. In the hands of such folklorists as Hans Naumann, the concept of "*gesunkenes Kulturgut*" lent itself to arguments that "the common folk do not create, they reproduce" and that folk songs are "botched art songs" ("*das Volk ein Kunstlied zersingt*").[126] Bogatyrev and Jakobson respond that "from a functional standpoint . . . the work of art outside of folklore and the work of art adopted by folklore are two fundamentally different facts." "Reproduction does not mean a passive appropriation," but involves "the selection and the transformation of the appropriated material."[127] Selection and transformation might appear to "botch" an appropriated art song, but only in the terms of the literate audience whom the folk song does not address. Bogatyrev and Jakobson energetically differentiate their model of cultural homeostasis from Naumann's quasi-Herderian imagination of the "primitive cultural community," "a kind of collective personality with one soul and one world-view which knows no individual expressions of human activity"; this leviathan alone, in the view of Naumann and his school, could be "the author of folklore, the author of collective creativity," just as "swallows' nests, bees' hives, and snails' shells are creations of genuine communal art."[128] To this the authors respond in conciliatory mode: it is risky, they say, to "deduce mentality directly from social expression . . . the unrestricted rule of the collective mentality is in no way a necessary prerequisite to collective creativity, even if such a mentality provides especially fertile soil for the most perfect realization of collective creativity. Even a culture permeated by individualism is not totally alien to collective creativity . . . [as in] anecdotes . . . rumors and gossip . . . legends, superstition and myth-making, social conventions and fashion."[129] The moderation of their language at this point contrasts strangely with their vigorous presentation of "preventive censure" in the earlier part of the essay. It is consistent, nonetheless, with the polemic in favor of a functionalist epistemology. Where Naumann and his Romantic predecessors operated with claims about essence, origin, and hierarchy, the young structuralists will answer that no hierarchy is static, that functions must incorporate variables, that identities are relative.

Nonetheless, when arguing for the social containment of cultural change the two Russian scholars had not made mild propositions. "Absolute dominance," "imperative," "indispensable," "everything rejected . . . simply does not exist," "abandon any attack": the curiously uncompromising, exception-free, and militarized language of Bogatyrev and Jakobson's description of oral culture suggests that the theoretical part of their essay is not concerned with describing empirically attested traditions, but with outlining a future of clear-cut distinctions. Only the purity of oral tradition can guarantee the clarity of their model (this contrasts, once more, with the acceptance of written-oral hybrids in their revision of Naumann's "appropriation" thesis). At many points, they concede that in a culture with writing, "preventive censorship" is apt to fail: a written document is never absolutely suppressed, it can always be hidden away in some private space to await its moment of public relevance.[130] But in an oral culture, according to Bogatyrev and Jakobson's model, information exists only in performance and in public; not only is there nowhere to hide it away, but revealing it would result in censure and the attempt to keep it hidden would result in its destruction through forgetting. Memory, in such a society, is only one generation deep. The "special type of participant in poetic creation" engineered by censorship reflects the community's ideal image back to it. A constant, anonymous feedback loop of censorship keeps the "community" identical to itself.

From the moment that someone can write a message and hide it away for a future reader, this cycle is disrupted. Writing preserves information that would otherwise be eliminated from the feedback loop. It interrupts the perfect mirroring of society and culture which "censorship" is intended to realize. It allows certain writers to aspire to "transform the demands of the milieu and to re-educate it through literature," the mission of every avant-garde known to history.[131] The redescription of oral tradition as a phenomenon analogous to *langue* thus leads Bogatyrev and Jakobson to exclude writing stringently from its concept.

No one in Germany, Russia, or Czechoslovakia in the first third of the twentieth century could use words like "Volk," "censorship," or "collective" without being aware of their public connotations. Four years after the publication of Jakobson and Bogatyrev's essay, the essay's implicit target, Hans Naumann, whose Nazi sympathies had long been in evidence, would join the Nazi Party and be chosen as rector of the University of Bonn. On May 10, 1933, Naumann supervised one of the first book burnings in Germany, titled *Aktion gegen den undeutschen Geist* (Demonstration Against the Un-German Spirit), delivering a lyrical oration in which, as the books were enveloped in the flames, he called on his hearers to burn what you have certainly not worshiped but what could seduce and threaten you and all of us. . . . If one book too many flies into the fire tonight, that does not do as much damage as if one too few were thrown in. . . . This fire is a symbol and shall also be a challenge to us, to clear our own hearts. . . . *Heil* then to new German writing! *Heil* to the highest *Führer*! *Heil Deutschland*![132]

Through this "symbol" and "challenge," the folklorist expressed the desire for authentic *primitive Gemeinschaftskultur*, for the power to remove from present consciousness "what could seduce and threaten you and all of us"—for what Bogatyrev and Jakobson named "preventive censorship"; the privative definition of orality in action. But under the conditions of book culture "clearing" the "hearts" of the *Volk* to that degree was unthinkable, and the ritual a futile symptom. To be effective, the fire would have had to purge Germany of all traces of any writing whatsoever.

2 / Writing as (One Form of) Notation

It is even more than likely that someone else would have done this before had it not been for the lack of the mechanical means: it has only been in the last few years that the science of electrical sound recording has given us an apparatus of such a sort that it can record songs of any length and in the large numbers needed before one can draw conclusions, and finally which can make records which are so good that the words on them can be accurately written down for the purposes of close study.

—MILMAN PARRY, 1935

To an ever greater degree the work of art reproduced becomes the work of art designed for reproducibility.

—WALTER BENJAMIN, 1936

Oral poetic theory is curiously obsessed with machines. Albert Lord defends Milman Parry's distributional study of the Homeric formula from the charge that it turns epic poetry into the mere whirring of gears:

> Are we not conceiving of the formula as a tool rather than as a living phenomenon of metrical language? . . . A style thus systematized by scholars on the foundation of analysis of texts is bound to appear very mechanical. Again we may turn to language itself for a useful parallel. . . . We find a special grammar within the grammar of the language, necessitated by the versification. The formulas are the phrases and clauses and sentences of this specialized poetic grammar. The speaker of this language, once he has mastered it, does not move any more mechanically within it than we do in ordinary speech. . . . Style is not really so mechanical as its systematization seems to imply.[1]

The accusation goes back to the first publication of Parry's theses. One reviewer balked: "Even admitting that 90 percent of a Homeric bard's performance consisted in mechanism, it still seems to me that allowance must be made for the poet's personality."[2] For most participants in the oral poetry debate, the line between the tradition and the individual poet must be displaced either by discovering flaws in Parry's theory of formula (which would allow partisans of Homer the individual artist to reclaim territory from determinism) or by reading passages in a way that will demonstrate the relevance and poetic intention, in other words the authorial design, of the words chosen. Homer was no machine, but a man; or (as a fallback position) Homer was a man who used a machine to achieve human ends. The consensus from which most debates begin is that a traditional, and by implication oral, method of poetic composition excludes individual design. Whatever is conceded to one side must be taken from the other. It is a rare scholar who, like Gregory Nagy, accepts the fullest (some would say the worst) implications of oral-traditional composition:

> Parry's work on the mechanics of Homeric diction has caused a serious problem of esthetics for generations of Hellenists reared on the classical approaches to the *Iliad* and *Odyssey*: how can compositions that have always seemed so deliberate and integral in their artistry result from a system of diction that is so mechanical—one might almost say automatic?[3]

The Epic Cyborg

Lord's protest resounds throughout *The Singer of Tales*, where the theme of the oral tradition as a living growth is insistent. Properly absorbed by gradual and organic learning processes in the singer's community, folk expression must be distinguished from the "meretricious virtues" of an art that "could be attained only through writing." Critics should know their limits: "Our greatest error is to attempt to make 'scientifically' rigid a phenomenon that is fluid." "One does not lead Proteus captive. To bind him is to destroy him."[4] In the longer perspective according to Lord, "Art appropriated the forms of oral narrative. But it is from the dynamic, life principle in myth, the wonder-working tale, that art derived its force."[5] This habit of setting living orality against mechanical literacy blocks the usual objection to Parry's statistical account of epic poetry, with the added facility of discouraging further analysis of the "phenomenon" Lord is describing. Its strategy is rhetorical chiasmus, a crossed substitution of properties: it is *because* the theory of oral composition suggested soulless mechanical production that Lord, and Parry at certain moments too, must defend it by claiming for it the opposite connotations of life, spontaneity, and soul. With that, writing, the scorned and scapegoated twin of orality, can (or must) be chased away, excluded, treated as a source of contamination. By thus insisting on its purity, oral-poetry theory has put itself into a false position and condemned itself to dogmatism and defensiveness.

In order to back out of this rhetorical cul-de-sac, we will need to re-verse the chiasmus and describe epic poets as animate machines and the transmission of oral poetry as a kind of intersubjective inscription that takes human beings as its writing tablets. (This task is already sketched out for us in the standard designation of Homeric epithets as "clichés" or stereotypes: if the epic bards did not write, they were expert in typogra-phy.) We will need, too, to take the measure of technological determinism in the longer history of oral poetics.

Ever since Parry first made the case for "oral-traditional poetry," cer-tain pieces of argument have come to rest in its foundations and proved difficult to dislodge. One is that poetry is either oral or literate. The in-terpreter attuned to the oral-poetry theory will have the job of determin-ing to which category a certain piece of poetry belongs. Second, "oral vs. written" is a zero-sum game. Discovering that a phrase or scene occurs more than once in Homer marks a point for the oral-traditionalist team; showing how a detail is irreplaceably right for its specific context marks one for the individualist-intentionalists.[6] Whatever is taken from the one side is given to the other. Third, oral poetry is said to be a "technique," but the explanation of this technique consists largely of enumerating the properties of another technique that oral poetry is said to lack.[7] (This other technique is, of course, writing.) Fourth, the desire to claim terri-tory for the newly discovered phenomenon incites critics to look aside from cases (the great majority of cases as they stand) in which the oral and the literate are mixed, interrelated, or overlaid. The most desirable informant will therefore be a singer innocent of letters and books, and performers capable of reading are suspected of inauthenticity.[8]

Such assumptions have become habitual in oral-traditional studies. Yet, as we know, the direction of the concept of the formula up to and includ-ing Parry's French theses had been toward a collective, systematic, gram-matical model of the "technical reproducibility" of poetry, hard to square with the later primordialism, vitalism, and exclusivism.[9] And Parry's re-sponse to a philological problem as old as Europe was systematic, statis-tical, and technological—utterly "American," in the sense that Bell Labs and the Model T were. His first publications laid out only the finely ad-justed parts of the Homeric machine, describing the formulaic language of epic as a "traditional" style, one that could not have been the creation of a single poet. In his English-language articles published soon after his return to the United States, where the traditional character of Homeric verse is made the premise for advancing the argument that the epics had been composed and transmitted orally, the reasoning is signposted with a rhetoric of the most straightforward technological determinism.

It must have been for *some* good reason that the poet, or poets, of the *Iliad* and the *Odyssey* kept to the formulas even when they, or he, had to use some of them very frequently. . . . What was this constraint? . . . There is *only one* need of this sort which can even be suggested—the necessity of making verses by the spoken word. This is a need which can be lifted from the poet *only* by writing, which *alone* allows the poet to leave his unfinished idea in the safe keeping of the paper which lies before him, while with whole unhurried mind he seeks along the ranges of his thought for the new group of words which his idea calls for.[10]

Since [the oral poet] is composing by word of mouth, he *must* go on without stopping from one phrase to the next. Since his poetry has being *only* in the course of his singing, and is *not* fixed on paper where it can show itself to him verse by verse, he *never* thinks of it critically phrase by phrase, but *only* faces the problem of its style when he is actually under the stress of singing.[11]

"Only—never—only": these logical connectors mask a wide gulf of un-considered alternatives. There is no middle ground in Parry's argument. Orality is defined by the absence of paper and writing: orality is nonwrit-ing, illiteracy, the lack of a temporal reserve. Condemned to live in the present tense and the linearity of "the course of his singing," the oral poet cannot afford the luxury of critical comparison: "he must go on." "Un-like the poets who wrote, [the oral singer] can put into verse only those ideas which are to be found in the phrases which are on his tongue. . . . At no time is he seeking words for an idea which has never before found expression, so that the question of originality in style means nothing to him."[12] The admirable "economy" of his system arises out of indigence rather than elegance. We can be sure

that Homer's style was oral. For there is a simple, almost too obvious, fact to show it: namely, that there is no memory of words save by the voice and the ear. . . . There is no real memory without sound. As a rule we are unable to recall a single phrase of the book we have read silently. The poet who is repeating his own phrase, or that of another, is doing so by ear. To deny this for any poet is to suppose impos-sible things . . . But when we come to Homer such a thing is beyond reason. . . . Homer could only have learned his formulas by hearing them spoken in the full voice of those poets to whom he listened in his childhood.[13]

Without the "necessity," there would be no purpose for the "technique": this comes to be the point of Parry's statistical and comparative studies.

The equivalences of paper with premeditation, and of hearing with immediacy, remain in the background of oral-formulaic theory. They have been there for a long time, in fact. "It has often occurred to me to doubt," says Rousseau,

> not only whether Homer knew how to write, but whether anyone wrote in his time. . . . If the *Iliad* had been written down, it would have been less sung, the rhapsodes would have been less sought-after and would not have grown in number. . . . Other poets wrote, only Homer sang, and these divine songs ceased to be heard with delight only when Europe became overrun with barbarians who dared to judge what they could not feel.[14]

Contrary to usual practice, Rousseau's "barbarians" are the literate, not the unlettered, hordes. (Luddites can be technological determinists too.) Robert Wood directly attributed Homer's poetic virtues to the oral medium in which he worked:

> When the sense was catched from the sound, and not deliberately collected from paper, simplicity and clearness were more necessary. Involved periods and an embarrassed style were not introduced until writing became more an art and labour supplied the place of genius . . . it was therefore an advantage to the Father of Poetry that he lived before the language of Compact and Art had so much prevailed over that of Nature and Truth.[15]

For Herder, it was obvious that Homer

> never sat himself down on a velvet cushion to write a heroic poem in twice twenty-four books according to the Aristotelian rules or, if the Muse so desired, against and beyond them. Rather he sang what he had heard, he depicted what he had seen and experienced in the flesh. . . . All artificial hindrances and word-labyrinths are foreign to the simple singer, he is always audible and for that reason always understandable.[16]

And so, in the middle of the twentieth century, Albert Lord continues the motif:

> We have exercised our imaginations and ingenuity in finding a kind of unity, individuality and originality in the Homeric poems that are irrelevant. Had Homer been interested in Aristotelian unity, he would not have been Homer, nor would he have composed the *Iliad* or *Odyssey*. An oral poet spins out a tale; he likes to ornament, if he has the

ability to do so, as Homer, of course, did. . . . The story is there, and Homer tells it to the end. He tells it fully and with a leisurely tempo, ever willing to linger and to tell another story that comes to his mind. And if the stories are apt, it is not because of a *preconceived* idea of structural unity which the singer is *self-consciously and laboriously* working out, but because *at the moment* when they occur to the poet in the telling of his tale he is so filled with his subject that the *natural* processes of association have brought to his mind a relevant tale.[17]

Homer ever "natural" and thus unfailingly "relevant," the modern critics doomed to discover only "self-conscious," "preconceived," and "irrelevant" ingenuity: for Lord as for his predecessors, medium is destiny.

A certain technological determinism can thus be laid at the door of those who have tried to uncover the specificities of oral tradition and oral literature. It is an inverted technological determinism, however, one that puts the *lack* of writing at the head of the conditions that dictate the properties of oral poets and their poems, the possible forms of epic consciousness. Having this lack in common, all preliterate texts are thought to be comparable or at least concurrent testimonies to the worldwide category of oral literature; and one expects that the common condition of orality should reveal itself in common properties of the texts constituting the category.[18] Correlatively, writing is said to create new and distinct conditions of possibility: new objects, new types of relation, and an altered consciousness for those who participate in it.[19]

Technological determinism is not a besetting sin or an error to be proscribed, save insofar as it leads to a neglect of the questions that most need to be asked. If only it led to consistently persuasive results, it would be a virtue. Technologies play a part in any human endeavor and influence the minds of those who narrate history, but they do not explain everything by themselves or relieve us of the necessity to look at the contingencies surrounding them. Technologies are determining and determined; to overlook the ways in which they are subject to chance and circumstance is to promote them to Promethean status.[20] How does a particular technique interact with the givens it operates on and the techniques it replaces? What adaptations do users make to it? And what sort of thing is the *absence* of a device? What is the necessity of a technology—how do needs make themselves felt and weigh on the forms of their resolution? Any study of orality and literacy will have to give a prominent place to the technologies of writing: the founding presupposition of the field is that the difference between orality and writing is significant. But what is the difference, and what does it signify? More than a contrast with literacy is required before we can hope to understand orality. Writing and the powers and values ascribed to it must be examined case by case. Is writing at bottom one thing "from China to Peru," or does it shatter into ad hoc local combinations of devices? The adoption of writing is often seen as the great watershed in the history of a society: Is that picture justified, or does it reduce to an aftereffect of the representational technologies to which historiography is most indebted? If techniques as such were adequate to explaining human activities, and any regime of technology were sufficient to understanding all the technological articulations that had gone before and into it, these would be idle questions.

But first: What kind of thing is writing? Is it different from recitation? If so, how?

"Word for Word"

As do X-rays, a technology reveals certain things by suppressing others. It imposes its mode of discovery. Writing enjoys this privilege: it is so much the standard in the immediate domain it covers that our usual term for an artist in words is "writer," and we rely on metaphors of inscription to imagine every kind of information retention, from historical awareness to genetic codes. But the more pervasive a technology, the more it is apt to become invisible, to take charge of the thinking of those seeking to think about it and school them in the selectiveness of its perceptions. Wood and Herder were so impressed with the figure of the bard who composed without pen and paper that they credited him with an utter spontaneity, as if to go without writing were to live entirely in the moment and to be free from dependence on any convention. (After transcribing part of an Eskimo dirge, Herder exclaims: "This Greenlander obeys the finest laws of the elegy form . . . and from whom did he learn them?"[21]—the point of his rhetorical question being that it is expected to remain without an answer). Thus writing comes to possess the whole territory of memory, custom, habit, learning—everything that is "programmed"—and the rest is spontaneity. In this, of course, the rediscoverers of orality followed writing's lead from beginning to end. They took a particular form of writing to be the condition sine qua non for the purposes served by literacy in their own societies, construed the absence of writing as the absence of everything for which writing has served as a means, and ignored the ways in which memory or habit might no less effectively "inscribe" precedents, customs, and phrases on the minds of individuals. It is as if observers had spent centuries describing jazz improvisers as defective performers in a classical tradition, and then, on noticing that they did not play from scores, erred on the side of attributing to them no preparation or forethought.

If writing is so successful at imposing its profile even on the orality that is said to be its negative, the attempt to draw a line—historical, psychological, technological, poetic—between oral and written tradition may simply falter. An example will show in what sense this is meant. In a passage setting out some of the decisive orientations of "orality" as a field of research, Albert Lord recognizes the need for a philology of oral diffusion that will break with the hallowed conventions and obsessions of text philology.

> Whereas the singer [of epic poetry] thinks of his song in terms of a flexible plan of themes, some of which are essential and some of which are not, we think of it as a given text which undergoes change from one singing to another. We are more aware of change than the singer is, because we have a concept of the fixity of the performance or of its recording on wire or tape or plastic or in writing. . . . We can say, then, that a song is the story about a given hero, but its expressed forms are multiple, and each of these expressed forms or tellings of the story is itself a separate song, in its own right, valid and authentic unto itself. . . . Our real difficulty arises from the fact that, unlike the oral poet, we are not accustomed to thinking in terms of fluidity. We find it difficult to grasp something that is multiform. It seems to us necessary to construct an ideal text or to seek an original, and we remain dissatisfied with an ever-changing phenomenon. . . . The truth of the matter is that our concept of "the original," of "the song," simply makes no sense in oral tradition.[22]

Read for its thematics, the passage lines up the characteristics of literacy and orality in two contrasting series: literacy establishes an original text in relation to which subsequent versions are more or less accurate, contain more or fewer "variants," whereas in orality, every performance of a song is equally "original," one of many "expressed forms" of a transmitted substance ("the story") that is never on show as such.[23] The "uniformity" expected of copyists in a literate tradition is scarcely imaginable in an oral tradition, for technical reasons whose social consequences have been drawn by many subsequent researchers.[24] The mode of existence proper to a work of verbal art in an oral tradition is "multiformity"— Lord's coinage, and a term that has traveled far.[25] But *we are more aware of change* than the singer is (my italics). Is it really a matter of awareness? Minna Skafte Jensen observes that dictation does not seem to have any special place in the singer's consciousness. . . . There is nothing to indicate that [Mirash Ndou, the Albanian epic singer whose performances Skafte Jensen recorded in 1974] felt that on that occasion he had made specially important versions of his songs, much less that he would try to repeat these special versions afterwards. This would be contradictory to his way of thinking, since to him a song *did not exist in ever-changing versions*; it was the same song, to be repeated whenever some kind of audience was interested.[26]

The hallmark of oral culture, its multiformity, becomes perceptible to observers with techniques for storing and comparing a plurality of performances (transcriptions, phonograph records, etc.); what the participants in that culture are conscious of is not multiplicity, but uniformity. Who then is "accustomed to thinking in terms of fluidity"? Apparently not "the oral poet." The overt thematics of the passage from Lord's *Singer of Tales* must, in some respects, be reversed. Although it looks as if writing introduces uniformity into a landscape of oral multiformity, in practice the multiform character of oral composition cannot readily be articulated without the use of some inscribing technology to capture the multiforms and present them for comparison. Literacy and orality thus do not constitute two separate worlds or eras that must be understood by applying separate standards to each, but a series of interrelated analytic protocols. The topography and chronology of the question cannot be dominated by a divide—the difference between oral and written mentalities, or the irruption of writing onto preliterate society—but rather appears as an overlay, which must be taken in its full complexity if it is to be understood at all.

As so often happens, the language of relativism conceals the mechanisms of relation. One of the most frequently enumerated characteristics of orality hinges on the difference between word-for-word memorization of texts and their recreation by the traditional means of formula, set verse, and theme. Albert Lord even went so far as to refuse to consider a body of traditional songs, transmitted with care for exact verbal fidelity among the unlettered Gilbert Islanders, as coming under the purview of oral literature: For him, such texts "[might] not be oral compositions, but rather written compositions without writing."[27] In this articulation we see that the empirical question of the origin of the texts (were they chanted or written, recopied or memorized?) has become secondary to the question of the "mentality" that produced them. For Lord, word-for-word replication always stood for "writing," whatever the means used, so that "rote memorization" was indicative of a "written mentality" even among people who did not use writing. Thus among oral performances there are those (the spontaneous, multiform ones) that are "truly oral," and those that are oral merely in fact. This surely involves the promotion of a specific case to the status of model for all cases, a contestable logical move (the "True Scotsman") as well as an undue restriction of the powers and techniques credited to the users of "orality." "Word-for-word" is a technical term that supposes a means of capturing the "words" and a purpose for which this or that definition of "word" has relevance. The question to ask is not: "do we have word-for-word transmission?" but "what type of transmission is considered word-for-word in this situation, and why?"

Singers have their decided opinions on fidelity. Asked by Nikola Vujnović, Milman Parry's assistant in 1934–35, whether he would ever shorten a song when pressed for time, Salih Ugljanin responded with vehemence:

> By Allah, you have to sing the whole song all the way through to the end, no matter what. . . . By Allah, there are shorter songs, but I can't leave one in the middle. . . . I can shorten it by telling it faster, by singing faster, but if I were to tell it slowly I can't.
> [Nikola:] I don't mean faster, but shorter. Would you leave any words out?
> [Salih:] By Allah, I can't leave anything out; that would spoil it. . . . It wouldn't be like the other one afterwards, *when they were put together*. They wouldn't agree.[28]

But Demail Zogić, who as proprietor of a Bosnian coffeehouse and occasional singer was particularly well placed to judge the differences among rhapsodes, had his own verdict on "leaving things out":

> [Nikola:] Tell me this, Demo. When two singers sing a song, you said last night that they can't sing the same song exactly alike. . . . Can that be?
> [Demail:] Yes. It is impossible to find two singers who can sing a song through clearly from beginning to end, but one will make a mistake, or will add something. . . . I prefer the man who leaves something out. . . . I always prefer to listen to him. The man who adds something thinks it up himself and adds it to the song, and it can't be true. As far as I'm concerned, let a singer sing it as it was. Then I like to listen to it, but I don't like under any circumstances to listen to the man who adds something to a song . . .

[Nikola:] In other words, as far as you are concerned, the song is the
   song, is that it?
[Demail:] That's it.[29]

Fidelity, as expressed by these South Slavic informants, refers not to
a prior exemplar but to an ideal of completeness, adequacy, and explic-
itness. The substance of the songs is historical, which accounts in part
for the demand that a version be "true" (true to type, taken as implying
historical accuracy). "Leaving things out" is a sign of careless practice,
just as is stopping halfway; "adding something" to an old song raises the
suspicion that the singer is vain or incompetent, though the ability to "or-
nament" a bare plot line with descriptions and incidents is much prized.
Asked if it is possible for a singer to repeat a song "word for word" after
hearing it only once, Demail Zogić insists that

> It's possible . . . I know from my own experience. When I was to-
>   gether with my brothers and had nothing to worry about, I would
>   hear a singer singing his song to the gusle, and after an hour I
>   would sing his whole song. I can't write. I would give every word
>   and not make a mistake on a single one. . . .
> [Nikola:] Was it the same song, word for word, and line for line?
> [Demail:] The same song, word for word, and line for line. I didn't
>   add a single line, and I didn't make a single mistake. . . .
> [Nikola:] Does a singer sing a song which he knows well . . . will he
>   sing it twice the same and sing every line?
> [Demail:] That is possible. If I were to live for twenty years, I would
>   sing the song which I sang for you here today just the same twenty
>   years from now, word for word.[30]

Seventeen years later, Demail Zogić made a second recording of the
song ("Bojicić Alija Rescues Alibey's Children") he had delivered on the
day of his conversation with Nikola Vojnović, and the two versions differ,
as might be expected, in manifold ways great and small.[31] It appears that
the recording machine has become a lie detector, and caught the singer
promising more than he can deliver. Was Demail Zogić exaggerating?
Or did he not share his questioner's meaning of the phrase "word for
word"? Lord's phrasing—"flexibility," "multiformity," "fluidity"—has the
ring of special pleading, as if to say that the unlettered singers should not
be held to the same standard as our copyists or lawyers, for their techniques are unable to fix the level of detail that we expect from writing or
magnetic tape. That is one way of being charitable; another way would be
to take Demail Zogić at his word and assert that his two recitals do cor-

respond "word for word," but in a sense of "word" that correlates with the purposes of Demail's technology for information storage and retrieval.[32] Gesemann noted that the Serbo-Croatian epic singers he studied a decade before Parry sang not from fixed texts, but from "schemata ... [that] have become to a certain degree teachable and learnable."[33] Would Demail's understanding of the term "word" be better rendered as "schema," "proposition," "event," "motif," or "speech act"? Under this second interpretation, the meaning of "word" is technologically situational, deriving from the devices used to capture "words," whatever they are. Rather than say that we have a *fixed* understanding of "words" in the one case and a *fluid* understanding of them in the other—which would mean that we had already assented to one definition of the "word" and one benchmark of "fixity" to cover both kinds of cases—we would require of ourselves a description of the ways in which the different technologies isolate and preserve the objects they both (perhaps ambiguously) call "words." The result is no longer a relativism in reference to a fixed standard, but, in recognition of the incongruity of different perspectives on fixity, different identifications of the thing that one repeats.

Before the application of the phonograph to collecting oral texts, the record shows little obsession with documenting "word for word" accuracy. Radloff's investigations among the Kara-Kirghiz in the 1870s led him to believe that

> every singer with some talent improvises his songs on the spur of the moment so that he is not even capable of reciting a song twice in completely the same manner. . . . The improvising singer sings without thinking about it; he sings only about the things he has always known when someone encourages him to sing; he sings like a speaker whose words come out of his mouth continuously. . . .
>
> A text that is not accurately written down and therefore not fixed is always in a fluid state and becomes something completely new in ten years. According to my experience, then, I hold it impossible that so enormous a work as Homer's poetry could have survived a decade had it not been written down.[34]

But this understanding of the inevitability of variation may correlate with Radloff's recording techniques. "The singer is not used to dictating so slowly that one can follow with a pen; therefore, he often loses the thread of the story and maneuvers himself into contradictions by omitting things. . . . Under these circumstances, the only thing left for me to do was to have a singer recite one episode to me first while I was taking notes about the development of the episode, and then I could proceed to the recording [in writing] when I was familiar with the content of the episode."[35] It sounds as if the act of taking down the text molds it into episodes and motifs—as if, once more, the transcribing technology does not stand apart from the performance it records, but affects its style and structure.

The "word for word" of writing is different from the "word for word" of oral recitation insofar as writing (and associated technologies) permits the two to be compared; to know what the shape of the relation would be if the "word for word" of oral recitation were taken as the standard, we would have to define the technologies of oral recitation otherwise than as a defective instance, or a happy absence, of writing. The unhappy consciousness of orality studies lies in their dependence on writing to delineate the object of their attention, that "nonwriting" that must be preserved against the inroads of even such forms of writing as "rote memorization." But that is only a further reason that we should refuse to see the relation of orality and writing as a "before" and an "after," two distinct historical spheres. Lord's description of the Gilbert Islanders' learning style as "written composition without writing" can be taken more positively, once it is admitted that writing is not the other whose exclusion founds the category of orality. If anything, as we have seen, the deployment of writing in certain specialized contexts permits orality and multiformity to emerge. Subjecting the various methods of information storage to questions about the process of the constitution of their objects improves the yield of our relativism, it seems, in respect of both specificity and overlay.

To summarize, then: writing, sound-recording technologies, oral-aural memorization, and—why not?—the use of knotted strings, petroglyphs, dances, or painted story scenes are all to be seen as rival systems of notation, each of which produces characteristic forms of inscription and reading.[36] The fact that some of these were developed with the purpose of capturing the inscriptional objects that we tend to recognize as relevant ("words" in the case of alphabetic texts; pitch, tempo, intonation, timbre, background noise and so forth in the case of tape-recorded performances; to which add facial expressions, gestures, and backgrounds in the case of cinema) does not guarantee that those objects are the right ones to capture. The actual relationships are apt to be far more complex and indeterminate. The oncologist interested in the properties of a tumor will rely on X-rays to supply certain kinds of information, cell cultures to provide others; it is a matter of deciding what to look for. The unstated privilege of "writing" in most discussions of orality appears through its fixing the criteria of relevance, accuracy, fixity, univocity, durability—in short, telling us what to look for (whether we find it or not).

What would a transcription system developed specifically for Homeric poetry look like? We might begin with the part of oral epic that most resembles, if we follow Albert Lord, "written composition without writing": the word-for-word reproduction of certain type-scenes. In dealing with them print technology is surely maladaptive, writing out every word of a recurrent scene, every time. In their very redundancy, type-scenes suggest a more economical representation of the structure of the epic document. For example, a five-line sequence describing serving maids as they bring bowls of water for hand washing and prepare for a meal (χέρνιβα δ' ἀμφίπολος προχόῳ φέρουσα . . .) recurs unchanged eight times in the course of the *Odyssey*. Rather than reproduce it in full each time, as a printed edition does, a hypertext edition would include the block of text once and for all in a lookup table and insert links to that section of the table at appropriate points in the text. Such cross-referencing would reduce the volume of data—by a factor of seven-eighths in this one instance. Thousands of other repetitions great and small could be registered in the same way.[37] It would be interesting to know how much bulk could be removed from an epic poem, without losing any of its information, by resorting to data-compression and data-expansion algorithms of this kind. It might also give us a rough image of the "competence" needed to produce our version of the *Odyssey* by differentiating among passages of pure routine, those that combine existing material in ways for which precedent exists, and those that, occurring but once in our corpus, would probably have had to be worked out from scratch. Just as the computational effort of a navigator is reduced by having on hand trigonometric tables and landmark coordinates, so the random-access memory of a reciter, one can imagine, meets regularly occurring events with regularly worded descriptions and stores up effort for the exceptional event that will need special verbal labor.

The enumeration of Homer's repetitions did not wait for the invention of hypertext. Arming scenes, assemblies, greetings, and combat scenes, just to name a few, have been analyzed into their constant and free elements.[38] The fact that these can be tabulated, and exceptional cases described as "the prototype, minus this and plus that," suggests that performers in the tradition learned such scenes algorithmically, one model covering a number of uses. The content of memory is not a "what" but a "how." "The finished singer will boast that he knows '*how to* saddle a horse,' or 'dress a hero,' or 'plan a battle' . . . Questioned further, he will explain that bad singers leave this out and leave that out, but a good singer knows *how* to put it back in, even though he has never heard that particular song."[39] The roster of typically occurring elements of a typical scene, transmitted from singer to singer by example, subserve the "how" or the art of that sort of scene. When stretches of text are memorized, it is in view of a purpose; they are strategic resources. "Dressing a hero" is a skill, something one learns how to do, that may be done better or worse. So, building from small events to great, is telling the *Iliad*. A singer's memory, we may infer, is teleological, acquiring resources in view of subjecting them to larger purposes. But the sort of writing we are used to calling "writing"—the notation of whole texts in "words" and "verses"—deals with the "what," not the "how," of speech, and that makes it an awkward and redundant notation system for the products of mnemonic technique.[40] The Oxford Classical Text on my table treats the χέρνιβα δ' ἀμφίπολος passage and others like it with a naïveté and literalism that a rhapsode would surely find distressing. A table of links, on the other hand, keeps one copy of each recurrent element and merely directs verbal traffic to it as needed, illuminating a structure of "hows" behind the "what" of print. Thus, an electronic device "remediates" a shortcoming in our ordinary writing and reading of formulaic texts.[41]

Every transcription system is selective and directive. Musical notation, for example, determines performance only in an approximate sense: whether it is as detailed as an orchestral score or as loose as a sequence of chord intervals written on a bar napkin, it always only lays out a subset of the actual properties of any future rendition as a basis from which performers start. Much must be supplied from tacit knowledge, habit, or interpretive choice, or made up on the spot; some things can never be known in advance. No one supposes that knowing how to write music is the same thing as musical knowledge, or that the product of one system of notation is more truly music than what another system notes. Western staff notation has a more loose and conventional connection to even Western music than seems to obtain, in the minds of most trained readers, between words and alphabetic notation.[42] The ease with which musicians and mathematicians can imagine alternative notation systems for their objects of concern is worth emulating. As is the freedom with which some musicians experiment with notations that are not necessarily realizable as sound.

"The map," as the saying goes, "is not the territory."[43] Yet mapping a territory makes it easier to do certain things with it. The forms of notation in use, for example, have much to do with the development of musical genres and their transmission to different places and publics. Ancient Chinese musical theory is much concerned with fixing the dimensions of the pitch-pipe that will serve as a constant and replicable standard for the beginning of the scale: without such an artifact, the classical texts suggest, the domain of music will be given over to an unbearable relativism.[44] It is hard to imagine that European polyphony could have survived for hundreds of years or been added to the common musical property of much of the world if the requisite techniques of inscription, including scoring and tuning, had not been able to fix many of its characteristics, just as much of what is valuable in jazz would have survived only as storytelling if the player piano and phonograph had not been on hand to preserve, spread, and transmit performance styles that few took the trouble to copy down in score.[45] Musical practice does not reduce to musical notation, but the two are nonetheless in a relationship in which each primes the other's development.[46] Such relationships cannot be predicted *ab initio*, but must be taken historically, as parts of a coevolutionary constellation change and elicit change in the other parts.

Neither the perfected, nor the defining, nor the inevitable form of notation for verbal texts, word-for-word writing selects some dimensions of a performance and ignores others; it foregrounds some and backgrounds others; and in this it is like every other notation. It has its objects, its saliences, its ways of shaping what it records. But the same must be said of other notation schemes. What are the other possible means of storage and retrieval of a complex, nonsingular object such as a verbal work of art? Can these devices be defined positively, rather than negatively as the absence of another, prevalent, scheme?

We might describe oral tradition as a poetic technology marked by collective composition, modularity, iterability, and virtuality. ("Marked," that is, in contrast with the poetic practices most familiar to us.) In *collective composition*, the right to determine the content of a performance is distributed widely throughout the community of performers; even where a norm exists, it does not exclude variation or improvement. *Modularity*: poems are combinations of preformed units that can be put together variously; any two different works in a tradition will tend to have many of these units in common. *Iterability*: a poem is not a final result but only one exemplar in a series of recitations, and to be preserved it must be recomposed again and again, modularly, by members of the collective. *Virtuality*: what is passed on and learned from poet to poet, if this is seen as occurring, is not the poem itself, a determinate series of words from beginning to end, but rather a recipe or strategy for making a poem that will answer to such-and-such a description. Conversely, no particular rendition of a poem exhausts the possibilities of the poem's tradition. The ambivalent relation of such traditions to the usual systems of notation lies in the fact that notation has been limited to recording the particular renditions, leaving the potential dimension of alternative realizations to be inferred. The oral is that virtual or ergodic register.

A formal or functional characterization of oral literature like that just proposed avoids tautology ("oral literature is literature produced and transmitted orally"), privative definition ("oral literature is literature made without writing"), and evaluation by preset standards ("oral transmission is inherently unreliable"). It defines oral tradition, rather, as a mode of production and circulation to which voice, orality, is incidental. Indeed many of the same features can be found in twentieth- and twenty-first-century literary avant-gardes.[47] With such a description in mind, we can reconceive literacy as well. Fixing a particular wording as definitive, closing the era of collective composition, making formulaic utterance redundant, reducing iterability to repetition, and restricting the scope of virtuality, the adoption of manuscript notation as the home for works of verbal art must have been as much a social change as a technological one. And not necessarily an advance.

Stitches in Time

In an example of the interplay (never identity) between a medium and the objects it produces, early print culture revealed properties of oral-formulaic poetry that neither recitation nor manuscript had paused over.

The fourth chapter of Ludolf Küster's previously mentioned *Historia critica Homeri* (1696) pauses over an apparently minor problem, the derivation of the word "rhapsode." Should *rhapsōidos* and *rhapsōidia* be connected to the verb *rhaptō* (meaning "to stitch"), or to the noun *rhabdos* (a staff or branch)? Two arguments are offered in favor of the first view. The rhapsodes might have been styled "stitchers of songs" by Pindar and Hesiod[48] because they were the first to collect and "sew" end to end the scattered bits of Homer's poetry,[49] or again because the rhapsodes used to improvise poems by "stitching" into new compositions preexisting Homeric lines irrespective of their earlier context.[50] Küster finds support for the first hypothesis in a number of Greek texts, chiefly a scholiast's note on Pindar, *Nemeans* ii. 2, where the rhapsodes are described as "drawing together the hitherto uncollected poetry of Homer . . . in a sort of linking or stitching."[51] Flavius Josephus, the Byzantine *Suda* lexicon, and Eustathius's commentary are cited in support of the argument that Homer left his work "neither assembled nor internally connected," so that it had to be ordered by later hands. Eustathius explains the practice of the rhapsodes thus: "Stitching together [συρράπτοντες] [their compositions] from parts taken freely from either of the two Homeric books [sc. the *Iliad* and the *Odyssey*], they got the name of 'rhapsodes' . . . . A 'rhapsody' is thus a poem stitched together [συρραφεῖσα] out of the two Homeric poems, according to a proposed theme."[52]

Such a technique of recomposition from fragments was still familiar to Küster's contemporaries. Such were "the highly elegant *Homerokentra*, or Homeric centons on the life of Christ"—odes and narratives on Christian themes, sometimes of surprising length, which medieval scholars used to form out of chains of Homeric quotations.[53] Is this what the rhapsodes did in their "stitching"? Küster is doubtful: what distinguished the rhapsodes from all other poets or wits is the fact that "they sang, not their own songs, but those of others" (*non sua, sed aliena, carmina cantare solerent*).[54] It seems that the boundary line of literary property is just about to become visible, for the *centones* contain nothing but Homeric verses although Homer is not considered their author; conversely, why are Petrus Candidus or the Empress Eudocia not considered rhapsodes?[55] With the start of the next paragraph, however, Küster abruptly drops this line of etymology and moves to accept the less troublesome derivation from *rhabdos*, that is, the reciter's emblematic laurel staff, long since offered by the poet Callimachus and the grammarian Dionysius Thrax.[56]

"They sang, not their own (*sua*) songs, but another's (*aliena*)"—the problem of defining a rhapsode and of describing his "stitching" comes to a head in the very sentence that is brought forth as the touchstone for distinguishing rhapsodes from their descendants the centon makers. For what is the exact measure of "one's own" and "Homer's own" in a poem that Homer is said to have left "not in one place or internally connected," so that later generations of rhapsodes either "drew it into a unity whenever they sang it" (according to Pindar's scholiast) or "stitched it together in whatever way they wished" (according to Eustathius)? *Homerokentra*, whose every line is a word-for-word citation of the primordial poet, are nonetheless not in Küster's mind the literary property of Homer; and yet the ancient rhapsodes who organized the component parts of Homeric poetry into new wholes enjoy no authorial status (except in the mode of blame). Ownership, in the matter of epic poetry, comes down to the difference between materials and labor, but later generations have no way of making that distinction clear-cut. Küster sees that the meaning of "rhapsode," if it is to occupy a place in an implicit series including "poet," "editor," and "publisher," rests on elusive bases; that may be his true reason for preferring to identify the rhapsode more simply as the bearer of a *rhabdos*.

Rhapsodes are, or ought to be, mere executants. For "with the passage of time, Homer's poetry became thoroughly corrupted by its rhapsodes, who often imported their own verses into it [*ut qui multos de suo versus ei infererent*]. This created the necessity for critics of emending Homer." The bad rhapsode tries to be a poet, disregarding the difference between "his own" and "another's" poetic property. In so doing, he creates his own antithesis and corrective in the person of the critic. Homer's moral right to the poetry that goes under his name must be asserted, for if (for example) the rhapsodes were thought to share in the creation of the poetry of Homer by their arrangement and unification of it, there would be that much less reason to object to the insertion of new verses in an already diverse and ill-defined corpus, and that much less conspicuous an "original glory" for critics to recover.[57]

In the history of texts, Küster sees the corruption of the authentic by the inauthentic. If criticism is the reversal of that process, Henri Estienne has bad news for critics. In his preface to a collection of post-Homeric writings including centons and parodies, Estienne asserts that these works fall under a regime of mixed, derivative ownership and cannot be definitively assigned to a single author. A parody of the *Iliad* or *Odyssey*, "although it is not ascribed to Homer, is nonetheless in some measure Homeric, in that it is artfully sewn together [*consarcinatum*] out of Homer's words, or perhaps from whole phrases taken from his verses."[58] The parodist's trick is to take over the wording of Homer's statements to express a meaning or situation at a great remove from the original one, and Estienne calls this hijacking "a kind of theft . . . but done in such a way that it can hardly be called stealing."[59] Parallel to Küster's unreliable rhapsodes (who foist on Homer what is not his own and thereby appropriate his name), Estienne's sticky-fingered parodists, like the composers of centons, force "another's" poems to serve "their own" meanings. Parody is an alienation of the poem from its author and from its prior integrity. But to trace the history of this alienation it will not be enough to go back to the first performance of a Homeric poem by someone other than Homer, for as Estienne observes with apparent wonderment, "we can say that Homer gave the very first example of parody, as when he copies his own verses and reuses the same lines (or lines for the most part made up of the very same words) on different occasions."[60] Homer's ownership of his own poetry—the assumption on which Küster built his justification for the critical enterprise—is here undermined by the image of a Homer who recurrently steals from his own pocket, turning his own best verses, through repetition, to parody. In Küster's account, the possibility of knowing Homer was challenged by the meddling of unreliable rhapsodes; in Estienne's, it seems that Homer has joined the rhapsodes and himself no longer knows the difference between "his own" and "theirs," between creation and repetition, between faithful transmission and burlesque.

Despite their differences in tone (one prim, one comic), these accounts of the relation among the Homeric corpus, its transmission, and Homer as its putative author have a common subject: they show something going wrong, an assumed understanding being broken. Estienne and Küster make the history of the Homeric corpus into a drama of alienation.[61] Text criticism presents itself as a form of intellectual property law (long before there was any such thing in the real-life republic of letters). But Homeric poetry is, in that case, an awkward experimental sample, since neither rhapsodes nor author seem to know the slightest thing about giving proper credit. They needed print to show them how to do that.

The self-similarity of the text and the moral identity of the author have a common origin here: writing as revealed through printing. We will have to resist the pull of centuries of ready-made debate and bracket the question of Homer's authorship: for to begin to ask, as is traditionally done, what in the Homeric poems is Homer's and what is not, and how we might ascertain the difference, is to adopt a frame of reference characteristic of manuscript and print cultures, bookselling, plagiarism, and forgery.[62] Within that frame, one could be as playful as Estienne or as schoolmasterly as Küster, but all within the same assumptions. We need to ask rather: Does the problem take the same shape under conditions of manuscript circulation or public recitation (these two media being susceptible to further subdivisions as necessary)? If there is a problem of ownership, is there an object to be owned, stolen, violated or restored? What is that object? Does it lend itself to playing the part of "property"? Or is the object known as "the poetry of Homer" ambiguous, with a Homeric performance, a hand-copied Homer, and the printed text of Homer supporting different social roles?

A profile of the object Küster and Estienne have in mind emerges from the verbal actions of which it is the target. "Homer's poetry," as whole or part, is "collected" and "woven" or "sewn together" (*colligere, syrrhaptein, contexere, consarcinare*); from its fragments new wholes can be "concocted" (*conficere*); conversely, pieces of it can be pried loose and stolen away (*furtum*), and the fabric of the whole can be "corrupted" by the "insertion" of foreign matter (*corrumpere, inferre*) or, happily, "mended" by the critic (*emendare, restituere*). These verbs are surely not only to be taken in a physical sense—nowhere is it claimed that "Homer's poetry" is a unique material object like the Mona Lisa or the Parthenon—but the relevance of the physical sense is nowhere in doubt: what has happened to "Homer's poetry" over the centuries is analogous to what might happen to a piece of cloth or a manuscript book. "Homer's poetry" is an object whose form and integrity reside in the continuity of a boundary or envelope that is under frequent threats of violation in the form of partial thefts and insertions. It has at least one dimension (its parts can be connected or disconnected) and possibly two (with edges that can be sewn together). Tragic for Küster, comic for Estienne, the violations nonetheless confer on the Homeric corpus an imaginary skin (with the qualities of outwardness, phenomenal appearance, and barrier function).[63]

The ancient sources of both discussions, however, were not so clear about the physics of the object and the actions in which it figures. A long citation will be necessary:

> [Pindar, Nemean Ode 2:] As the Homeridai,
> bards, chiefly begin their stitched songs
> with an invocation to Zeus, so this man . . .
> [Scholia (selected):] Just as the Homeridai {and the sons of
> Homer, as the poets believe}, when they rhapsodized {independently
> from the tradition, καταχρηστικῶς} {and composed certain songs},
> are said to have begun with [the invocation of] Zeus, so this man
> [Timodēmos] has laid a foundation [for his future victories] . . . by
> coming in first at the Nemean games [which were held in the honor
> of Zeus]. . . .[64]
> Anciently the name *Homeridai* ['sons of Homer'] was applied
> to those who claimed descent from Homer and sang his poetry in
> succession [ἐκ διαδοχῆς]; afterwards, it included rhapsodes who no
> longer traced their birth to Homer. Well-known among them were
> those in Kynaithos's circle: they are said to have made many [lines?
> episodes? books?] of the epics and inserted them into the poetry
> of Homer [οὕς φασι πολλὰ τῶν ἐπῶν ποιήσαντας ἐμβαλεῖν εἰς τὴν
> Ὁμήρου ποίησιν]. Kynaithos was a Chian by birth: he wrote down
> the 'Hymn to Apollo' from the poems ascribed [ἐπιγραφομένων] to
> Homer and dedicated it [ἀνατέθεικεν, as when offering an ex-voto in
> a temple] to him. . . .
> Another explanation. Some say that the original of 'rhapsodes' is
> 'rhabdodes,' from the custom of traversing [διεξιέναι] epics 'with a
> staff'—that is, Homer's epics. Callimachus says: 'And the tale woven
> over the *rhabdos* . . . I sing continuously as I received it.'[65]
> Others say that when the poetry of Homer had not yet been col-
> lected into one, but was scattered about in pieces and [recited] dif-
> ferently here and there, the rhapsodes, whenever they sang it [ὁπότε
> ῥαψῷδοιεν αὐτήν], aligned the parts with one another through a kind

of linking and stitching [εἱρμῷ τινὶ καὶ ῥάφῃ παραπλήσιον ποίειν] as they drew it into one.

Yet others say that at first, because the poetry was transmitted in separate parts, each of the competitors sang whichever part he liked, and at that time the [reciters] were called ἀρνῳδοί [*arnôidoi*, "lamb-singers"], from the lamb offered as a prize to the winners. But later, whichever of the two poems was introduced [ἑκατήρας τῆς ποιήσεως εἰσενεχθείσης] [as the subject of the competition], the contestants, as if patching [οἷον ἀκουμένους][66] the parts toward one another, advanced upon the whole poem; from that time they were addressed as "rhapsodes" [song-stitchers]. . . .

Another explanation. The 'Homeridai' at first were the children of Homer, and later the rhabdodes around Kynaithos. For these rhabdodes memorized and declaimed [ἀπήγγελλον] the poetry of Homer, which had been dispersed; and they harmed it [ἐλυμήναντο] very much.[67]

This amalgam of late-antique reports on the institutions of centuries past gives a good image of the kitchen where much of our knowledge of classical literature is made. Many of the terms used are open to confusion: For example, when it is said that the Homeridai "sang [Homer's] poetry *ek diadokhēs*" (successively), the phrase can mean that the tradition was passed down from generation to generation, or that their performances were organized as relays, one singer taking up where another left off. In fact, both meanings of *diadokhē* find confirmation in the scholiasts' stories, which might leave one doubting the accuracy of either.

The term "Homeric poetry" is contextually ambiguous too, in the same ways that Küster tried but failed to ignore. Sometimes it seems best to take it as referring to the whole cycle of epic poems, including but not limited to the *Iliad* and *Odyssey*, as when it is said that Kynaithos and his disciples, "having made [ποιήσαντας] many [parts] of songs [πολλὰ τῶν ἐπῶν], inserted them into the poetry of Homer." (The scale of *epē* is not clear: whole epics, verses, or epic episodes?) At other times, when delimited with a number, "either of the two poems" [ἑκατήρας τῆς ποιήσεως] and "the whole body of poetry" [τὴν σύμπασαν ποίησιν] designate something that—at least for the author of the scholion—has taken definite shape and borders. But the same author admits that the "whole poem" yet contained gaps that called for supplementation. Was the "patching" done by the rhapsodes merely a matter of execution? The making of poems is likened to "weaving," an act that ought logically to precede the "stitching" and combination of woven pieces—but Callimachus's verse specifies that the Homeric *muthos* was "woven over the rhabdos," the emblem of the rhapsode's office, thus eroding a possible and tempting distinction between a primary "weaver" and a secondary corps of "stitchers." The verb used in one scholion to designate the selection of a subject for competitive recitation, εἰσφέρω, "import," closely resembles the word used in another scholion to describe the rhapsodes' "insertions" of new material into the epics (ἐμβάλλω): if both mean roughly "to bring [something] into a bounded space," their mutual resemblance suggests a continuity, not a difference, between the legitimate institution of song-contests and the dubious additions by latecomer reciters. The paragraph that has the rhapsodes "inserting" segments into the Homeric corpus (indicating a considerable degree of fluidity in the contents of that corpus) also refers to the poems in the language of a literary, even bookish age as "ascribed" to Homer, and treats as noteworthy—perhaps even unprecedented— Kynaithos's act of writing down one segment from Homer's store and transforming it into an offering. We hear that the rhapsodes "wronged" or "dishonored" the poetry, with a verb (λυμαίνομαι) often used in Greek to designate abuse to the body or the reputation, and applied at least once to an actor's "murdering" his speeches—but the exact nature of the injury to Homer we do not know.[68] (The word *lumainomai* could be taken as implying that the relation between Homer and his rhapsodes was conceived as analogous to that between dramatic author and actors in later centuries—a tempting but anachronistic analogy for scholars of the late classical period.) In sum, the scholiasts tell us little about the shape of "Homer's poetry" in the long period between its creation and the first explicitly mentioned standard texts.

If the name "Homer" was a question, the terms from antiquity that frame possible answers to the question are indefinite as well: the "poetry," its "composition" and "transmission," the "poems" and their "parts," be these "genuine" or "spurious": each of these keywords admits more than one possible scope, and their combinations are multiply enigmatic. Competing understandings of the preclassical past and changes in the reception of Homer over the long period of classical commentary doubtless account for some of this fluctuation. Clearly the scholiasts (and their readers in late antiquity and Byzantium) do not imagine "Homeric poetry" in the time of the rhapsodes simply or solely as a book. Along with their artifactual (clothlike, skinlike) properties, the epics have many properties of events or spaces. "*Whenever* [ὁπότε] the rhapsodes sang it" they did certain things to it—pulled the parts together, filled in the gaps—and the result of their actions was something to be "gone through" (διεξειμί, "traverse" as well as "relate, narrate") or "proclaimed" (ἀπαγγέλλω, like a law or a truce).

Although the scholiasts' narrations clearly imply generic boundaries enabling hearers to discriminate "Homeric poetry" from other types of poetry, to rank competitors, and to judge differences between successive performances (their "insertions," "patchings," and "advances"), we would need more evidence than is given to establish the existence of a model or standard against which a particular recital might be checked. A definite form of "the two poems" is implied by the story of how the parts came to make up those paired wholes, but that may be a phantasm achieved by reading backward from the possession of written texts. No definite measure is given for the "nearness" (παραπλήσιον, says the scholion) of the parts as they are progressively "aligned," leaving open the possibility that the rhapsodes could have gone on forever inserting new microepisodes into the general subjects of the *Iliad* and *Odyssey*; and this possibility is arrested only by our retrospective certainty about the contents of the two poems.

If we discount the specter of a book, the event-aspect of "Homeric poetry" comes to the fore.[69] In the scholiasts' story of the emergence of the poems as complete and connected wholes, "Homeric poetry" is the subject of ritualized contests structured, like games or drinking competitions, by rules and protocols covering occasion, theme, turn-taking, entry and exit. The rules, to extend this analogy, would establish the nature of the contest, but not its outcome. Putting aside as anticipatory the idea that the object of the contest was to produce the most accurate chanted reproduction of an already existing text, or even the most dramatically affecting version of such a text (either of which alternatives would make rhapsodes nothing more than actors), we might locate the antagonism of the contest in the struggle to make the best version; then the rules would touch on required or permitted features of poetic content, but leave much in the realization of epic texts up to the ingenuity of the contestants. The scholiasts, then, can be read as talking about a set of objects and practices at least partially distinct from those which their early modern readers understood them to be talking about.

Küster's standards for authorship—the skinlike integrity of Homer's poetic corpus and the difference between originator and transmitter—cannot be formulated clearly for a milieu where the literary heritage is "patched up" anew on every occasion of its recital, and physical and institutional barriers are lacking to separate one's own contribution from a predecessor's.[70] Küster's critical mission and Estienne's humor both derive from expectations based on the properties of the book. Those expectations do not easily survive transposition to a different technology of reproduction (for example, the singing competition) that produces differently configured objects. The point is not to accuse Küster and Estienne of an inability to recognize the dynamics of an oral tradition. It is rather to bring to light their involvement in a technology that has become invisible to them. Authorship and criticism both exist as aftereffects of determinate techniques of transmission—writing (of a highly specific type: word-for-word transcription, which is not the only form of writing), reading, copying, collating. Küster's criticism could not be critical about its own dependency on these techniques; it had to read the ancient sources as if from the beginning Homer were already an author and the Homeric poems, settled texts. Hence the discomfort with the "stitching" of the rhapsode.

Does this mean that the "Homeric question" evaporates from the moment we cease assuming that Homer wrote books? Or that we have only to forget about writing to learn that "many, if not most of the questions we were asking, were not the right ones to ask"?[71] We must reckon with the epochs of media, but not expect them to answer all questions or even to tell us which questions are unanswerable. A given technology enables a set of consequences: thus the technology that is responsible for the durable, continuous, dimensional, multipliable object "text" also makes possible a new set of relations to that object (authorship, recitership, interpolation, forgery, destruction, and so forth). It is also possible for an imagination to be constrained by that technology's set of possibilities, to the point of recognizing no alternatives to them (thus Küster and Estienne). That is an often unconscious form of technological determinism: a particular state of the art defines the true and the false, the authentic and the inauthentic, the accurate and the inaccurate, without tagging these determinations as derived from a certain experimental setup. Early testimonies of the telephone speak with awe of its exact reproduction of a living voice, but it is clear that auditory standards have changed since 1876. Likewise, observers of the first moving pictures were so taken with their faithfulness to life, when we notice rather their jerkiness, their lack of color and sound, their flatness, their static camera. What impressed early viewers was the *increase* in realism, by comparison with still photographs and magic lanterns; our senses are attuned to a different set of illusion-making devices. "Realism circa 1775" is not the same thing as "realism circa 1850," "realism circa 1930," and so forth. Each has its effective context and range of alternatives.[72] The added time variable and the relationships it refers to serve as a brake on the impulse to narrate the history of realistic representation as continuous and end-driven.

Jonathan Sterne reminds us that "sound fidelity is much more about faith in the social function and organization of machines than it is about the relation of a sound to its 'source.' . . . Sound reproduction—from its very beginnings—always implied social relations among people, machines, practices, and sounds."[73] Technologies of communication do not simply work on objects, but (in various ways) reproduce them and frame them, that is, mediate between their physical and social identities. However, the truth-claims of a technology belong to an environment containing other technologies and their histories. A technology makes possible certain claims—for recording technologies such as writing or photography, these will most often be claims about an unprecedented degree of accuracy. The substance of these claims does not easily transfer to earlier or parallel technologies. Thus the kinds of fidelity and the sorts of information preserved by sound recording differ from those attained by unassisted writing. The artifices may be compared in certain respects, but the phonograph (which grew, like the telephone, out of attempts to fashion an improved writing or "visible speech") effects a change in the definition of "fidelity."[74] Every significantly new recording technique does the same. As Estienne and Küster write in the age of advancing print, they deal with variants, forgeries, and misattributions, and quite naturally they treat the authenticity of the Homeric poems as another case of a kind with which they are familiar. If they pause to think about a technology of memorization and repetition, all they see is its potential for error, just as we are unable to ignore the echoes, fadeouts, and scratches on old recordings. In short, they write as if the transmission of dependable texts, in their sense of "dependable," had been a possibility from the beginning, and as if the technology they had come to rely on were the standard-setter for fidelity in general.

The interpretation of the characteristic objects of one technology in the terms proper to another is a clear case of imaginative constraint by Whiggish history. Analogous to it is the assumption that the historical sequence of technological modes of production is cumulative and transparent overall (thus, in the case of Homeric origins, the assumption that whatever is produced by memory and oral tradition is *also* produced and reproduced by manuscript, printing, sound recording, electronic storage, and so on).[75] As a corrective, we need to notice the ways in which the designated objects of each technology are not identical, do not play identical social roles or bear the same properties. "Homeric poetry" could not be the same thing for Küster and for the rhapsodes, even assuming (against all probability) that in some case the words conveyed by a manuscript Homer and an orally performed Homer were identical.[76] We must, I think, take stock of technological determination in order to ward off technological determinism. By observing the ways in which our imaginations are constrained by an available set of means, we may hope to arrive at a better understanding the differently ordered possibilities onto which technologies open.

Nothing is more common than to think of a successful technology as initiating an "era" (a horizon of possibilities that it, or an imaginary perfected version of itself, is seen as fulfilling): thus we have the age of sail, print culture, the nuclear era. But there is a covert determinism in the idea of technological "eras," if they are thought to imply that the shift to a new technology happens instantly, uniformly, and for purely technical reasons. A sentence from Marshall McLuhan economically illustrates this wishful narrative pattern: describing televisual communications as "the cosmic membrane that has been snapped round the globe by the electric dilation of our various senses" permits the devices, all by themselves, to create the conditions and manifest the implications of global villagehood in a single, decisive, present-perfect-tense "snap."[77] What McLuhan wrote in the grip of his powerful vision needs rather to be questioned, the membrane analyzed into its constituent parts and relays, the instantaneous "snap" subdivided into beginnings, middles, and ends.

The consequences of any material or intellectual breakthrough are easy to enumerate in hindsight, hard to frame as a general law. But how the story is told will depend on how much stress is laid on the particular technology chosen, its meshings with other technologies, ambient conditions, and actors (where known). Various technological Whiggisms, romanticisms, or Luddisms arise from strategic choices of emphasis and level of detail in tracing these causal networks. Students of literature may find it useful to conceive of a technological regime as a language, with its semantic domain (field of effective objects) and syntactic rules (set of possible combinations among objects and actors). Among languages, translation is often awkward, never perfect, but the claim that "there is no word for X" in a certain language is always to be distrusted. There may be no obvious choice of a single word to represent a single word in some other language, but it is the translator who decides that an equivalent is needed, and settling on "no word" may only demonstrate the desire to find no word adequate. So, too, for the claim "there was no technology for X." Determining the limits of available means supposes defining the purposes in question. If we say that, before writing was applied to the traditional poetry of ancient Greece, "there was no technology for fixing editions or authorship," are we conceiving of "editions" and "authorship" so transparently in the terms of script that the statement reduces to a tautology, or have we at least tried to imagine a translation of those terms into a context that would not assume what we know as writing? Conversely, to say that "the phonograph opened the possibility of X" is not only to name an event in history but also to mediate among the languages proper to several phases of technical consciousness (the semantic field prior to the phonograph; that field plus the phonograph; the field, including the phonograph and its derived technologies, from which we speak today, and so forth).

The task, then, is to conceive of technologies (oral recitation, handwriting, and so forth) and their affordances within a social frame and to assess the ease of translation from that social frame into others. A certain version of the experienced world of the ancient Homeric reciters, audiences, transcribers, and readers appears in the poems themselves: the idealized portraits of the bards Phemios and Demodokos, the simile likening Odysseus to a harpist, Achilles's practice of singing heroic lays when at ease in his tent, Helen's mimicry of the voices of the Achaean warriors' wives[78]; but apart from these well-known passages, any portion of the poems having to do with the mechanics of communication can be taken as self-reflexive. Mechanics, that is: inscribing, transmitting, recognizing; reporting messages, seeing to it that orders are carried out. What do they tell us about composition and transmission? The *Iliad* contains scenes of "writing," if we are willing to extend that name to any use of the word *graphein* (γραφέω, "to wound lightly, to scratch, to sketch," and thus later "to write"). It contains scenes of "reading" (γιγνώσκω) as well, as when heroes throw lots consisting of marked stones and claim their prize by "recognizing" the lot that comes out victorious.[79] (The famous Bellerophon incident could be a negative version of this martial "recognition."[80]) The suspense of the *Odyssey*'s plot hinges on the recognition—alternatively the "reading"—of its hero, marked graphically on the thigh by an old hunting accident and thus made available for conclusive identification by any person "literate" in the reference of that sign.[81] (As the bearer of a written message, Odysseus is therefore not speaking out of character when he says, scandalously for a native of an oral culture, that a repeated tale is hateful to him.[82]) In these and many other scenes of transmission, it is hardly a matter of "fluidity," hardly optional, whether the hero is recognized, whether Zeus's orders are followed, or whether Athena is successful in "bringing to mind" her favorite hero in the assembly of gods (minus the hostile Poseidon). As Françoise Létoublon puts it in speaking of the *Iliad*,

> the words of Zeus, pronounced in the presence of messengers whose task it is to transmit them, repeated by them to the designated recipient, and sometimes repeated in front of or by other persons as well, are a capital element in the text's organization. . . . Thus Zeus supervises and directs not only the deeds of men, but also the reflection and representation of these deeds in human memory, through the epic text.[83]

Nothing could matter more to the epic than the success of inscription, so understood.

If writing, in a broad sense, matters to the epic, any clarity we might achieve by excluding it will be short-lived. But (not to expend all our cultural relativism on distant epochs) do we know what "writing" meant to the theorists of orality? How many kinds of "writing" are there? How many of them might be hidden within the current category of "nonwriting"? With a farewell to Whigs and Luddites alike, we embark on a journey through the storehouse of abandoned media.

3 / Autography

As frequencies and new rhythms will have to be indicated on the score, our current notation will be inadequate. The new notation will probably be seismographic.

—EDGARD VARÈSE

The standard-setting function of writing in the discussion of oral litera-ture is double. First, its absence defines what is "oral"; and second, as a technology, it exemplifies the standard of precision in comparison with which oral texts are said to wobble: writing is the image of language be-cause it captures "word for word," "letter for letter," what speakers say. But is this the right or the highest task for writing? Do we even know what writing is?

These discussions assume a certain kind of writing: alphabetic or pho-netic writing, as it is conventionally called. In the Greek world, alphabetic writing appears at roughly the time when ancient historians first begin to speak of transcriptions of the *Iliad* and *Odyssey*; one opinion has it that writing down the epics was the purpose for which the Greek alphabet was evolved.[1] If writing is not the image of spoken sounds, what is it—ideo-graphic? The category is less than clear.[2] Scholars in search of an "other" writing usually contrast alphabetic writing with the systems of ancient Mesopotamia, South America, or East Asia. These examples, though, are far from exhausting the potential reach of the concept.

The Inscribing Ear

Transcription, as we saw, makes the difference between two recitals of the same oral text being "the same" or merely "similar"—a distinction that seems lost on people who do not write. For some other purposes, the identity captured by writing is still too gross. "When, in a lecture, we hear several repetitions of the word *Gentlemen!*, we have the feeling that each time the same expression is in play, and yet variations in pace and intonation give it, in the successive passages, appreciable phonetic differences."[3] By calling our attention to variations in delivery, the linguist adjusts, momentarily, our sense of the threshold between noise and information. So do all technologies of notation. Sound recording preserves differences that transliteration lets escape. Punctuation and paragraphing give a speech a semantic architecture that might elude the listener. Alphabetic writing selects for the features and the degree of fineness that correspond to its structure; a writing differently structured would select for different ones; and why should people who do not write be "not writing" one set of features rather than another?

When Friedrich August Wolf commanded the would-be Homeric scholar to cast away all memory of books, ink and paper, the noting-down by hand of spoken discourse was a familiar practice. Its limitations were well known too. In the alphabetized cultures of Europe, sermons, debates, court proceedings and the like were taken down with a degree of precision relative to the purposes for which the transcription was made. Thucydides reproduces lengthy speeches of delegations, leaders and even hostages, but no one assumes that their immediacy is much more than rhetorical *inventio* on the historian's part.[4] A medieval interrogatory, like those used in Le Roy Ladurie's *Montaillou* to reconstruct the lives of thirteenth-century Occitan peasants, could be written in Latin and in indirect discourse; for the heresy hunters, what mattered was detecting erroneous beliefs, not the exact turns of phrase used by peasant men and women. (That the transcribers' skill preserved echoes of those turns of phrase is a happy accident.) In a democracy, the precise nuances of political speech came in for intense scrutiny by partisans, adversaries, and ideological hacks. The newspapers of revolutionary France or the American republic often consist of little more than transcribed speeches. But these were often touched up after the fact, or even invented by the journalists who knew the speaker's party line and their paper's attitude toward it.[5] In the best of cases, the hand had trouble keeping up with the mouth, and even shorthand failed when several people spoke at once. A gap between speech and print was apparent to any reader of Wolf who was paying attention in 1795, though when a medium is as pervasive as writing was then, its users must have been disinclined to notice any such gaps. Indeed, "the success of all media depends at some level on inattention or 'blindness' to the media technologies themselves (and all of their supporting protocols) in favor of attention to the phenomena, 'the content,' that they represent."[6]

Alphabetic writing was too slow and too loose-grained, according to inventors focused on the gaps. Alexander John Ellis, writing in 1845, scornfully renamed English spelling "English heterography" and its printing "heterotypy."[7] Of course, any alphabet rests on a convention.

> But it may be asked: will the same characters recall [the same sounds] to another person, who uses the same alphabet, and who has acquired the knowledge of its use, in the manner already indicated? Of this we can be by no means sure, because we are ignorant of this person's constitution and therefore cannot possibly state whether the same external causes will produce the same effect upon him as upon the first person.[8]

Ellis proposes to limit this variability by enlarging the Roman alphabet with diacriticals and compounds. It is not sure that these refinements will make writing a faithful and unchanging portrait of sound—after all, "the impossibility of there being any standard for pronunciation" means that "there are as many different vowel systems as there are speaking individuals in existence."[9] Ellis's innovation is to reconceive the act of reading as a cause-and-effect dynamic, an instance of playback with the reader being a medium driven by the written input: "The term 'Alphabet of Nature,' we should apply to a series of symbols representing certain Mechanical Conditions requisite for the production of the sensations termed spoken sounds, so that those conditions being fulfilled, the same set of sensations in the same order may be produced in any individual."[10] Ellis's ambitions are expressed by his rival in the field of "phonography," as the perfecting of alphabets was then called, Alexander Melville Bell. Bell's "Visible Speech" (1867) was to

> embody the whole classification of sounds, and make each element of speech shew in its symbol the position of its sound in the organic scale . . . . The correlation of the Sounds and Symbols rendered the latter SELF-INTERPRETING to those who possessed the key to the symbolism, and so converted the UNIVERSAL ALPHABET, which had been the object of the designer, into a real VISIBLE SPEECH;—the latter constituting, in fact, a new Science,—adapted for the use of all mankind![11]

Its benefits would include teaching the illiterate, blind, deaf, and dumb to read; the communication of the exact sounds of foreign languages to learners; setting a standard of pronunciation for each language; "prevention and removal of defects and impediments of speech"; telegraphic communication; "study, comparison and preservation of fast-disappearing dialects, and the universal tracing of the affinities of words; . . . speedy diffusion of the language of a mother country through the most widely separated colonies; . . . the world-wide communication of any specific sounds with absolute uniformity; and, consequently, the possible construction and establishment of a universal language."[12] As the *Athenaeum* admiringly put it, "A great many efforts have been made to spell words: but the system before us spells spelling."[13] In theatrical demonstrations, Bell showed audiences how far his invention outstripped existing alphabets. Alexander John Ellis himself testifies: "Mr. Bell wrote down my queer and purposely-exaggerated pronunciations and mispronunciations, and delicate distinctions, in such a manner that his sons, not having heard them, [but reading from a 'Visible Speech' transcription,] so uttered them as to surprise me by the extremely correct echo of my own voice . . . accent, tone, drawl, brevity, indistinctness, were all reproduced with surprising accuracy."[14] This must have struck the audience as something like the miraculous mechanical "phonography" that Edison would bring forth a decade later.

In 1857, a printer's assistant named Edouard-Léon Scott de Martinville submitted to the Academy of Sciences in Paris a sealed packet containing the history of his efforts to fashion "an apparatus that would reproduce through a graphic trace the most delicate details of the motion of sound waves," resulting in a "natural stenography."[15] The apparatus consisted of a cone-shaped receiver terminating at its small end with a thin diaphragm to which was attached a stylus of boar's bristle. This stylus, shuddering in response to the vibrations of speech sounds uttered into the cone, inscribed a wavy line on a rotating cylinder coated with lampblack. The resulting traces were *phonautography*: "the self-writing of sound."

> Our current writing expresses but one only of the modes according to which the voice represents thought; it is suited to representing nothing but articulation. Natural writing or stenography, of which here are the first rudiments, returns the rhythm, the expression thereof: it is a function of tonality, of intensity, of timbre, of measure. For this reason it is called to play a new and unforeseen role in the relations of intellectual life; it will be living speech; our manual or printed calligraphy is nothing but dead speech.[16]

Scott includes in his missives to the Academy of Sciences several examples of vibratory lines traced by his apparatus.

Noted on the margins of the lampblack paper rolls are the sound phe-

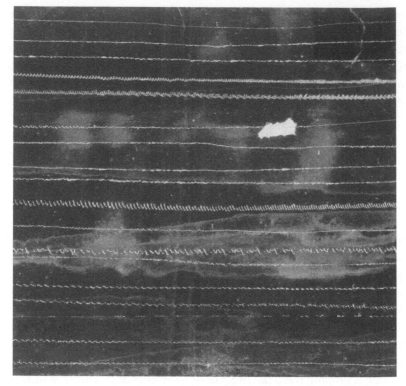

FIGURE 1. Phonoautographic writing of the human voice (detail). From Scott, *Brevet d'invention* (1857). By kind permission of the Institut National de la Propriété Industrielle, Paris, France.

nomena they transcribe: screams, cries, song, notes played on a cornet, and a murderous soliloquy from Ducis's adaptation of Shakespeare's *Othello*:

> . . . Dans leur rage cruelle,
> Nos lions du désert, sous leurs antres brûlants,
> Déchirent quelquefois les voyageurs tremblants.
> Il vaudrait mieux pour lui que leur faim dévorante
> Dispersât les lambeaux de sa chair palpitante,
> Que de tomber vivant dans mes terribles mains.[17]

The violence of many of these examples is explained by the difficulty of getting the stylus to respond to a subtler stimulus. Nonetheless the paper rolls record inflections that typography and musical notation could not have rendered with like precision: "[The] figures are large when the sound is intense, microscopic if it is very weak, spread out if it is low, squeezed together if it is high, of a regular and straightforward pattern if the timbre is pure, uneven and somewhat shaky if it is bad or muted."[18] These mappings of sound as visual pattern—what Nietzsche, mistakenly, would seize on as his metaphor for delusive sensory "metaphors"—were for Scott a way of "forcing nature herself to constitute a general written language of all sounds."[19] "Nature herself" performed the transcription both in virtue of general physical laws and because Scott had devised his apparatus as an "artificial ear." Like his device, the eardrum received vibrations from outside and inscribed them, in some mysterious fashion, on the brain, stimulating perception and leaving memory traces. The paradoxical formulation with which Pierre Jean Rousselot later designated Scott's invention—"the inscribing ear"—made evident the addition it made to the existing repertoire of the senses, for the ear had always been thought of as a receptive organ, not an active one.[20]

The traces inscribed by Scott's mechanical ear could not be played back, as could the lines etched by the later phonautographic machines of Cros, Edison, and Berliner. Scott's bristle operated a translation of sound into a visual analogue, reducing the three-dimensional orientation space of listening to the single dimension of graphic time. The idea of inversing the causal chain, of making the tracings stir the stylus and communicate its movements to a diaphragm that would render them as sound, had not occurred to Scott or Koenig, his instrument maker.[21] "Only the graphic inscription of sounds and the analysis it made possible concerned them."[22] Scott's device was purely an improvement on writing technologies. The one result he produced that could be read or played back was a bastardized or comic variant—a sheet of cursive calligraphy that borrowed the size and spacing of letters from the curves traced by the phonautograph, adding to the representation of words customary in alphabetic writing a representation of volume and emphasis.

Scott's device directly represented (wrote) physical processes unfolding in time and space, giving them a linear figure that could be examined for its qualities: wider or narrower oscillations, regularity or irregularity, tightness or looseness. Writing, it showed, did not necessarily consist of a finite set of distinct letters that combined in series to evoke meaningful words. But Scott's phonautographic inscriptions could not be read, despite a superficial resemblance with extremely hurried cursive.[23] The natural causality of its production made it an analog language, a silent literature that contained no differences, only variations (mostly of intensity). It was, additionally, the transcription not of a repeatable ideality (such as is the writing down of any letter, word, or sentence), but the trace

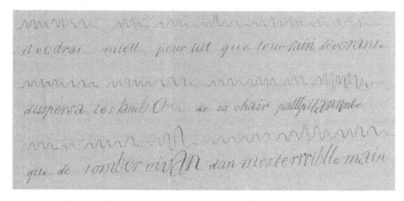

FIGURE 2. Quasi-phonoautographic writing (detail). From Scott, *Certificat d'addition*, 1859 (INPI ms. 31470). By kind permission of the Institut National de la Propriété Industrielle, Paris, France.

of an individual event with all its infinitesimal particularities of timbre, pace, ambient noise, and so forth.[24] Habituated as we are to phonographic playback, we take for granted the possibility of mechanical repetition of the individual sounding event (the singing of "Au clair de la lune" one moonlit night, replayable in the daytime and for any number of future days and nights), but Scott and his hearers stood just short of that technical horizon.[25] The phonautogram captures everything that is not repeatable in Saussure's example of the repeated word "Gentlemen!" Nonvocal, once-off, analogue, asemantic: What could possibly be the use of such a writing, in the terms for which alphabetic writing had served so long as a standard?

Scott's transformation of the auditory into the graphic was, we would be tempted to think, no Promethean resurrection but a dead end, a one-way translation. In the late twentieth century, however, following the invention of the laser disk, sound reproduction technology became predominantly optical. Physicists with an interest in early sound developed a device that scanned the inscribed surfaces of older recording media (lampblack paper, foil, wax cylinders, shellac disks) and read them, not with a physical needle but with a virtual stylus. For this optical technology there was no difference between the traces made by the speaking gramophone and those left by the mute phonoautogram: all were traces of vibration and all could now be played back. Thus, a parenthesis of adjournment opened in 1860 closed in 2008, when the 1860 sounds, sounds that had never been heard since being initially emitted, were liberated from their wobbling visual tracks. We can now hear someone, apparently

Léon Scott de Martinville, singing a stanza of "Au Clair de la Lune" in 1857, a record that Scott himself never succeeded in playing back.[26]

Scott's was one of a series of nineteenth-century machines that redefined writing in terms that corresponded to new investigations in physics, medicine, physiology, and linguistics. Essential to carrying out these investigations were machines that tracked movement and rendered it as lines or curves on a plane surface: the family of kymographs, developed by Thomas Young, Carl Ludwig, Etienne-Jules Marey, and others. In these machines, a writing point was propelled by the energy of some external process and made to represent the stages of that process by moving across a space defined by minutely subdivided time and some other variable of interest.[27] Ludwig's kymograph of 1846 represented variations in blood pressure as oscillations in a line traced on a revolving cylinder; Helmholtz's myograph of 1850 registered the timing and intensity of muscular twitches; Scott's 1857 phonautograph put the same principle to work on a much smaller time scale, with a needle that registered the impacts of successive sound waves on a membrane.[28]

These inscribing devices conveyed their impulses to a visual language that had been built up and refined since Descartes's analytic geometry. Consider the weight of a child, recorded at regular intervals from birth to age three. With the x-axis divided into segments corresponding to weeks or months, and the y-axis labeled in pounds or kilograms, each dated act of measurement will be plotted as a data point. Once a series of such points has been laid down, a line or curve can be drawn to estimate the growth process between data points. More than one such curve can be plotted on the same sheet, to enable correlations between rates of increase of height and weight, comparisons between boys and girls, growth statistics of different populations, and so forth. For some two hundred years we have been accustomed to scrutinize these quantitative hieroglyphics for quantities, rises and falls, rates of change, surpluses, deficits, regularities and irregularities.[29] Starting in the late eighteenth century, they entered service in economics (Playfair), meteorology (Young), ballistics (Obenheim), medicine (Lorain), public health (Bowditch), railway planning, cartography, engineering, physiology, and even linguistics. Graphic methods were the Big Data of their time. Their widespread use in France seems to have been a result of the teaching at the École Polytechnique, whose first director, Gaspard Monge, was a specialist in descriptive and analytic geometry.[30] Their best-known exponent, Etienne-Jules Marey, also a *polytechnicien*, applied the methods of geometry and engineering to the life sciences.

Marey's lectures *On Movement in the Functions of Life* (1868) specify the conditions of graphic observation.

Every phenomenon realizes itself in an act that requires a certain amount of time for completion. Thus an evaporating liquid takes a longer or shorter time to disappear entirely; a heated body reaches its maximum temperature sooner or later, etc. . . . in such a way that if we divide the whole duration of the phenomenon into more or less short segments of time, we will observe that the state of the body will be different in each of these phases. For the complete understanding of a phenomenon, it must be supposed that we know the state of that body at every instant during its change of state.

The more closely the successive observations follow each other, the more complete will be our understanding of the phenomenon under study. In graphic representation, we must then compile the greatest possible number of successive notations, each one corresponding to the state of the phenomenon at the moment of one observation.[31]

To understand "movement," then, we must divide time into ever smaller units and perform ever more meticulous observations of the x-axis variables. An illness, like a symphony, is composed of thousands of events, some simultaneous, some coordinated, some conditioned on others. Hailing polyphonic musical notation as the earliest means of precisely correlating events in time, Marey insists that "the introduction of the notion of synchronicity, or non-synchronicity, between two acts is of supreme importance. . . . In natural sciences, the mission of the graphic method is above all to bring out the relations among phenomena."[32] "Today, my conviction is complete: virtually every objective phenomenon, that is, everything that is a change in states of matter, can be studied by this method; not to use it, limiting oneself to mere observation and the inadequate testimony of the senses, would be voluntarily to commit oneself to error."[33]

Marey's *La méthode graphique* of 1878 proclaims the general applicability of graphing: "wherever the progress of a phenomenon is to be shown, [the graphic method] translates its phases with a clarity that eludes language."[34] This is partly a matter of the perceptual logic of the graphic method. "The graphic method is both a mode of expression and a means of research."[35] As a "mode of expression," it takes information in digital form—numerical tables—and not only renders it as points on a two-dimensional surface, which would simply amount to a translation, but fills in the transitions between data points with lines or curves, recreating the continuity that characterizes changes in temperature, growth, pressure, or the like. Even if the curves are added ex post facto, the resulting visual form aligns with analog experience and thus is perceived as intuitive, natural, believable, to a degree that perhaps "eludes language." We read them, at any rate, in analog fashion, assessing them in terms of more or less, faster or slower, and not in the this-or-that ("digital") manner appropriate to the letter or the word. Marey predicts that "one day physics textbooks will be nothing more than atlases of graphic tables containing curves superimposed and compared in various ways. Through these, we may grasp relations that we cannot even suspect today."[36] Graphic inscribing devices augmented human capacities, giving them "new senses."[37]

As a "means of research," the graphic method sought data that were immediately graphic. Such data were furnished by the use of inscribing devices like the kymograph, the seismograph, or the phonautograph. These allowed the investigator to "obtain with no effort at all a curve drawn by the self-inscribing phenomenon itself," rendering gradual analog events with continuous analog traces.[38] Marey's first great success was a sphygmograph that rendered the patient's pulse rhythms as a jagged wave on a turning drum. His machine resolved to a far finer level of detail than the traditional technique of medical palpation, as well as producing a retrospectively examinable public record on paper. With pulse, respiration, muscle movement, and nerve action among its first domains, nothing in physiology could resist the "inscription of forces," as Marey called it, on coordinates of space and time. The building block of Marey's investigations was the "tambour," a flattened drum of aluminum with intake and output tubes and a top side consisting of a rubber membrane. Pressure exerted on the membrane by the phenomenon to be studied could be redirected to a lever or to another tambour commanding a stylus. (The tambour closest to the subject under study was the "exploratory tambour"; the one in contact with the graphing paper was the "inscribing tambour.") Or force might be redirected to a lever, or to another tambour, by the action of air in rubber tubes.

Several processes simultaneously underway—for example, heart rate, breathing, and muscular contractions—could each transmit impulses to a dedicated tambour, which would relay them to one of several styluses inscribing parallel columns of data onto a roll of paper marked in increments of time. The tambours produce the first draft of a description of the event—be it the circulation of the blood, the change of seasons, the gait of horses or running soldiers—next to which Marey's verbal explanations occupy the subservient role of commentary.[39] The physiologist's job is to read the writing produced by the inscribing drums and to follow it with more writing.

Marey's aims were analytic. His chronophotographic studies of galloping horses and flying birds are famous precursors of the cinema, but

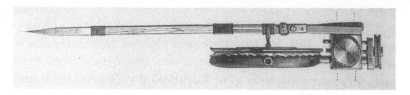

FIGURE 3. Marey tambour with inscribing pen. From Marey, *La méthode graphique*, 138.

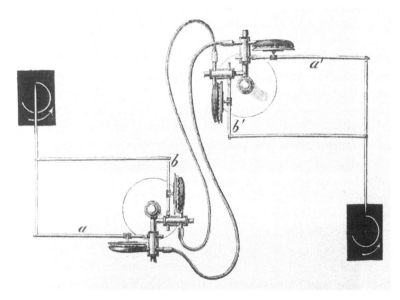

FIGURE 4. Four Marey tambours arranged to form a pantograph for graphic inscription (top view). From Marey, *La méthode graphique*, 132.

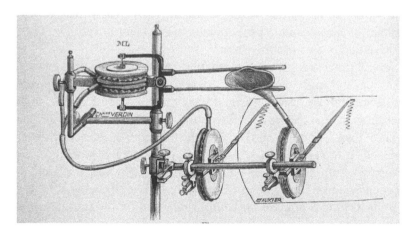

FIGURE 5. Registration devices for lip movement and voice vibrations linked to a cylinder by Marey tambours. From Rousselot, *Principes de phonétique expérimentale*, 1:92.

Marey apparently never thought of projecting them into space as illusions of reconstituted horses and birds. The purpose was to separate and clarify the different successive states, not to confuse them again by the persistence of vision.[40] Marey's skepticism about the senses is balanced by his investment in the "new senses" conferred on us by autographic writing machines. As with Scott, we must consider the implications of the fact that the first editions of the great mimetic technologies, sound recording and motion recording, were not directed toward playback but toward observation and transcription. Phonographs and cinema projectors address our *old* senses, achieving their "persistence" despite advances in analysis. Epistemologically, they represent a backsliding. We who are surrounded day and night by recorded sound and moving images find it difficult to see in Scott's or Marey's inventions anything but imperfect anticipations of the gramophone or the cinema, and thereby miss their medium-specificity. Hence the insistence, here, on characterizing them under the retronym "writing": it is less misleading to treat them as new developments of an old technology than to see them as defective versions of technologies that did not yet exist.

"Speech Is a Movement"

Pour durer, il faut donc se confier à des rythmes, c'est-à-dire à des systèmes d'instants.

GASTON BACHELARD

These new means of writing and the critique of writing coincided in the work of the Abbé Rousselot.

Born in a village of southwest-central France that had sometimes been attached to the province of the Angoumois, sometimes to the Limousin, Rousselot began his studies in his home region, obtained the baccalaureate (exceptionally for a pious boy, at a time when the Church looked askance on the French state and its educational institutions), and took holy orders, but found more opportunities for advancement in Paris than in the diocese of Angoulême.[41] A good Latinist, familiar with his local patois and with standard French, Rousselot, at age thirty-three, was driven to do field research by others' unreflective use of the alphabet.

To try a new path and exploit new materials: this is the idea that came to me in 1879, thanks to some bad books of Romance philology that I had stumbled upon. I was shocked to see them attending to the transformations of the letter rather than of the sound (for which the letter is a mere symbol), and, instead of studying a dead letter, I had the idea of studying living speech.[42]

He thought of writing a doctoral thesis on a subject then under discussion: the exact placement of the border between the *langue d'oc* and the *langue d'oïl* (the two major dialect families of French).

> The *Geographical Researches on the Langue d'Oc—Langue d'Oïl Boundary*, by Tourtoulon and Bringuier, had just indicated to me the interest of my patois, which straddles the line dividing the dialects of North and South. Trusting in M. de Tourtoulon, then, I began to study the sub-dialect of the Marches, under which is classified the patois that I have spoken since childhood, and I began to explore the region assigned to this dialect, going from village to village and questioning, with the help of the parish priests, people who had been born there from parents also native to the region, noting all the differences I encountered going from Saint-Claud (Charente) to Ids (on the far side of Montluçon): walking in pursuit of a limit that ceaselessly receded before me. The information I was able to gather took me as far as the Monts de la Madeleine, and there I stopped.
>
> From this first expedition I brought back ideas that were no longer those of M. de Tourtoulon . . . and something more valuable still, the habit of observation.[43]

"A limit that ceaselessly receded before me"—as an effect of sharpened perception and scholarly scruple, the line that had oriented the researches of his predecessors Tourtoulon and Bringuier—a line drawn on a map included in their pamphlet—dissolved before the investigator who tried to walk it. Tourtoulon and Bringuier's investigation of 1876 launched the period of systematic dialect research in France. Their purpose was to substantiate a division, dating from the Middle Ages, of France into southern (*langue d'oc*) and northern (*langue d'oïl*) regions; inasmuch as their research was sponsored by the Society for the Study of Romance Languages, based in Montpellier and established by the Félibrige group of Provençal writers and scholars, it could hardly have called the existence of that North-South division into doubt. To do so would have been to abandon the cause of a revived *langue d'oc* culture. Rejecting "the theory of the fusion of languages," the study's authors sought "to trace the line of demarcation with mathematical precision wherever possible."[44] This assertion of regional specificity represented one of the main options in dialect research, the other being an atomistic geography of countless centers each affecting their immediate neighbors. Tourtoulon and Bringuier's map drew a response from Gaston Paris, who preferred to see in the dialects of France, rather than a field divided in two by a "wall," a gently varying but single "tapestry"—a terminology that for southern scholars only echoed generations of Parisian, Jacobin centralism.[45] The "limit" expressed a rejection of the standard, national language. The *langue d'oc* activists' desire to trace a sharp line caused them to treat "mixed" dialects as unstable jargons of isolated words having no linguistic authenticity. Rousselot, however, retracing their steps, found the evidence for both clear boundaries and inorganic dialects wanting.[46] His story (as reported by himself) then takes a providential turn:

> I fell sick after returning from this trip, and I was obliged to stay in my family. . . . My mother became the subject of my study, and during more than three months, I had no better occupation than to espy her slightest utterances.
>
> I had never observed her previously, believing that my way of speaking, which I had picked up entirely from her, was identical with hers; but before long I discovered my mistake, and I was quickly convinced that the *geographical* study of patois must be completed by its *genealogical* study.[47]
>
> It was . . . after two months of careful observations around the Plateau Central, across the regions of Angloulême, Poitou, the Marche, and the Bourbonnais, that I first became aware of the differences between my mother's speech and my own. That was a discovery for me; and since then, each time that I have been able to assemble the members of a single family, the grandparents, the fathers, the mothers, the children, or people from neighboring villages, the same spectacle presents itself: a number of steps in phonetic evolution, staggered in close order.[48]

Rousselot had not only to notice the differences between his pronunciations and his mother's, but more significantly, to make of them something more than individual differences. For as long as there has been awareness of dialects, they have been thought to correlate with region: the linguistic atlas is a form that comes naturally to this field of research. Two villages close together on the map should be expected to have similar linguistic habits. A noticeable difference between the pronunciations of two so closely related people, coming after months of collecting information about the infinitesimal differences between the patois spoken in one village after another, presented Rousselot with a new variable. Language changed with time, as everyone doing linguistics knew; the most advanced scholars of language had declared that phonetic change admitted no exceptions and was "blind," that is, not driven by or toward a purpose.[49] But to see language changing, infinitesimally but audibly, in the speakers of a single household meant that it was deceptive to take even "the patois of Cellefrouin" (pop. 1,720 in 1886) as a coherent object of study. From now on, in Rousselot's methodology, the unit described by the dialectologist would be the patois spoken by a particular individual, born in such a year, whose parents, born in such a year and such a year, had been brought up in village A and on farm B. A "dialect" was a generalization among divergent behaviors. Every individual presented variations; geography and genealogy would put these variations in a systematic context of change—a tendency to drop a consonant here, sharpen a vowel there, presaged or exacerbated in the speech of other speakers placed farther along the dimensions of space or time.

Rousselot's thesis samples the speech of more than three hundred named individuals, for most of whom dwelling place, date of birth, some rudiments of genealogy and the date of the interview are given. For example: "*Les Pradelières*. M. Dubois, 43 years old, and his two sons, aged 20 and 13 (1887); his wife is from Mas-Dieu. I saw M. Dubois again in 1890; three children aged 12, 10 and 9 (1890). 184 inhabitants."[50] Summarized, the speech of these individuals attested, in his account, to waves of phonetic modification advancing gradually across wide swaths of territory.

[In the patois of Cellefrouin and environs] the *l'* [*l mouillé*, palatized *l*] tends to reduce to a *y*. This reduction has already occurred in the upper valley for final *l'*, and it has been followed by the complete omission of the *y*. . . . This evolution has stopped at the very limits of the commune of Cellefrouin. Today a new evolution affects the *l'*, whatever its origin, and its conditions allow us to observe it step by step and to note its stages. . . . The change makes itself known at Ventouse in two imported formulas of politeness, *o pyezi* ("*au plaisir!*") and *pyet-i* ("*plait-il?*"), and in Cellefrouin in a native word, frequently used, *pyo* (=flat), as of the generation of 1847. The change is complete with the generation of 1859 (Françoise Neuville). This new pronunciation seemed to be an impediment requiring correction. But nature does not readily run backwards. . . .

Despite moving into a place where the *l'* was not yet disturbed, my family carried the seeds of the change, and although we were born outside of Cellefrouin, my sisters and I were still subject to the influences that were apparent there. I (1846) still have the intact *l'*. But my sister Marie-Louise (1850) retains the *l'* only after gutturals, though she is able to pronounce it in all positions; Juliette (1852) has lost the facility of pronouncing it except in post-guttural position, and always renders it as *y*.

At Cellefrouin, all the generations posterior to 1859 have completely lost the *l'*.[51]

The very definition of patois makes it regional, but Rousselot's attention to small differences locates it in the individual, the farm or village, and the particular moment of time, seen as hosts of a wider development moving through the countryside.

> We have seen phonetic movements start from a determined point, gradually ascend the valley without being interrupted by administrative divisions of territory, propagating themselves at the most active centers, beginning with the most commonly used words, making their arrival known beforehand in remote places, hindered or hastened by the arrival of foreign elements in the population, snatching up children in their cradles and leaving the old undisturbed, but sometimes carrying off people of ripe age who get in step after making a voluntary and deliberate change, now charging ahead, now moving inch by inch, sometimes even falling back from conquered territory and beginning again, until the day that they become permanent, blotting out all discrepancies, just as if they had met no obstacle and had triumphed straightaway.
>
> It is like a sea submerging its banks.[52]

The three hundred or more speakers consulted in Rousselot's research were therefore participants in a drama, rather than simply informants. Their way of saying "rabbit," for example, characterized them as being among the avant-garde or among the resistance, relative to their place on the map and calendar.

Rousselot admits personal choice and overall determinism. Speakers may choose to affiliate themselves with other speakers of a regional patois, they may select certain sounds and avoid others, but in the long run "the instrument betrays the will. Evolution is unconscious." Change is "progressive as regards sound" and "progressive as regards place."[53] Children are most exposed to it, because they are least in control of their vocal apparatus and most susceptible to the organic motor of phonetic change, "a sort of anemia, a gradual and transitory enfeeblement of the nervous centers that terminate in the [vocal] muscles where [phonetic] evolution is seated."[54]

The drama of linguistic change, involving thousands of people mostly unknown to one another and behavioral trends so large yet so subtle that

lapin	Dinan	St-Carné	Meillac	Corseul	Quessoy	Mon-contour	Plénée-Jugon	Cesson
lapĕ	*y —*	—	*l i*	*ĕʲ*	*ãŋ*	*ĕ̃ŋ*	*ẽ*

C'est-à-dire :

	—	*yapĕ*	*yapẽ*	*lapi*	*lapĕʲ*	*lapãŋ*	*lapĕŋ*	*lapĕ*

FIGURE 6. Variants for the patois word equivalent to *lapin*, arranged by place of collection. From Rousselot, *Principes*, 44.

<table>
<tr><td>

ĕ júr kŏ y avĭ ún ŏm ĕ
ún fĕm k ăvyĭ sĕt afĭ.
lĕ pŭ jĕn, k ătĭ grŏ kŭm
rĕ, săpĕlăv l pĭ păsĕ.
y ătyĭ mălĕrŭ, mălĕrŭ
kŭm lĭ pyĕrĕ : ĭ n'ăvyĭ,
bŭn jă! pŭ d kĭt
pŭ ă mĭjă. ĕ sĕr kĕ lŏm
ĕ lĭ fĕm ătyĭ ă s ĕ̃fă, lŭ
pĭ̃ sŭ lŭ lădĭ̃ : « k vŭ-tŭ,
mŭ pŏv fĕm, kĕ dĭsĭ lŏm,
fŏb kĕ năjă lŭ părdr :
ĭn pŏ pŭ lŭ vĕr sŭfrĭ pŭ
lŏŭă. »

</td><td>

Un jour ça y avait un homme et
une femme qui avaient sept enfants.
Le plus jeune, qui était gros comme
rien , s'appelait le Petit Poucet.
Ils étaient malheureux, malheureux
comme les pierres : ils n'avaient,
bonne gent! pas *de quitte* (même de)
pain à manger. Un soir que l'homme
et la femme étaient à se chauffer, les
pieds sur les landiers : « Que veux-tu,
ma pauvre femme, que dit l'homme,
faut bien que nous allions les perdre :
je ne puis pas les voir souffrir plus
longtemps. »

</td></tr>
</table>

FIGURE 7. The beginning of "Le petit Poucet" (Tom Thumb), in patois and in French translation. From Rousselot, *Modifications*, 129.

they tend to go unnoticed, comes down to a question of muscle tone. The appeal to physiology is the token of Marey's method, becoming in Rousselot's hands a matter of necessity.

Thirty or forty sound features observed through selective interviews with three hundred subjects in twenty or thirty locations, over a time range amounting to a hundred years (supplemented by written documents going back another seven hundred): with attention to detail on this level, the usual techniques of transcription were bound to prove inadequate. How to register a fugitive phenomenon in all its variants? Like all dialectologists, Rousselot added diacritical signs to the alphabet in an effort to make transcriptions exact.

But "the ear does not hear everything, and we cannot assign a value to everything it does hear. . . . The scale of sounds is not the same for all, and we lack a fixed note, a reference-point, to serve as basis for our evaluations."[55] How could an alphabetic transcription mark the difference between Rousselet's and his mother's pronunciations of, say, "lapin"? We could imagine a notation that would render the first as *lapinJP* and the second as *lapinM*, but this device would have no information value; it would simply label two individual events without characterizing them.

Alphabets are not made for that. They deal in the typical, the repeatable; below a certain level of generality they cease to apply. Frustrated in his efforts to express in writing the kinds of distinctions on which his dialect survey rested, Rousselot consulted the medievalist Gaston Paris, who suggested using laboratory instruments.

Ten years earlier, a committee of the Paris Linguistic Society had launched a joint project with Marey investigating the muscular basis of vowel and consonant articulation. Using several "exploring tambours," they had demonstrated the interdependent action of the larynx, lips, and lungs and showed that the "same" consonant was produced differently in different consonant clusters. (The action of the tongue eluded their inge-

```
              ĕ   j----ă----r   k----ŏ       y----ă----v----i
      ┌ 1. 400  440.460.520   ·340                  320  480
L. A. │ 2. 520  600           440.440 380 400 360 420.440
      └ 3. 480  520.520       400        360     360
      ┌ 2. 280  240           300         240    240.240
      │ 3. 320  320                               320
L. E. │ 4. 400  380                               400.400
      │ 5. 320  280                               280
      └ 6. 320                       300          280

              ă--------n      ŏ------m   ĕ   ŭn   fœm
      ┌ 1. 440  360
L. A. │ 2. 400.400  360       320                320  400
      └ 3.          360
      ┌           ┌ 600(N)
      │ 2. 320    │ 150      320 280.240 280 360 240.400
      │ 3.                   320 320 320 320  320
L. E. │ 4.                                    400
      │ 5. 280               320 320 320 320 320.320
      └ 6. 300               320 320 320 320  300
```

FIGURE 8. A durational analysis (in milliseconds) of speech sounds in the dialect version of "Le petit Poucet." From Rousselot, *Modifications*, 130.

FIGURE 9. An analysis of speech melody in the dialect version of "Le petit Poucet." From Rousselot, *Modifications*, 133.

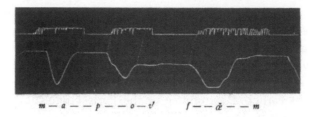

FIGURE 10. A fragment of "Le petit Poucet" as represented by the "speech analyzer": *"Ma pauvre femme."* The upper line traces the vibrations of the larynx, the lower the opening and closing of the lips. From Rousselot, *Principes*, 1:509.

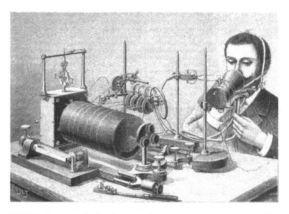

FIGURE 11. "Abbé Rousselot's device for the inscription of speech." From Brunot, "L'inscription de la parole," *La Nature* 998 (16 July 1892), 97.

nuity for the time being.) Whatever alphabetic writing might suggest, the *m* of *mp* was a different physical event from the *m* of *pm*. "Juxtaposition alters consonants, a most important fact for comparative linguistics."[56] Marey's recording devices would allow Rousselot to give numerical and visual form to the fine distinctions he was pursuing.

Working closely with the instrument maker Charles Verdin, Rousselot extended the repertoire of Marey's laboratory. His future research would be based on the results generated by a polygraphic device designated as the "inscripteur de la parole," combining a chronometer with instruments for recording lip movements, nasal pressure, the vibrations of the larynx, the movements of the tongue, and so forth, all these factors to be traced on a rotating drum.

Rousselot echoes Marey in his appreciation of the graphic method, which allows us to accomplish the often-cherished and apparently utopian idea of a universal alphabet, the phonographic alphabet, in which each articulation would be represented with the general characteristics that form its unity and also show it in the modifications deriving either from its juxtapositions or from dialects, which constitute its individuality. In this new alphabet, each letter would be composed, not of a more or less complex set of lines and dots with a value set by convention, nor of abstract formulas, but of the very traces produced by the corresponding articulation.[57]

In the narration of the outset of his career Rousselot performed the ritual opposition of the "dead letter" and the "living speech," but his version of orality is far from imposing a rejection of writing. "Bad books of Romance philology" pursue a trivial object, the recurrences of a letter over time, and proclaim that to be historical linguistics. With what he calls "the very traces produced by the corresponding articulation," Rousselot simply selects a different ontology of the events of speech. With his inscribing technology, speech can now be captured as an individual physiological event. What is speech? According to some, it is "the airy sign of thought, and thereby of the soul, which teaches, preaches, exhorts, prays, praises, loves; by which individuality is made manifest in life; nearly palpable for the blind, impossible to describe because too shifting and diverse . . . this untouchable, invisible voice, the most immaterial thing on earth, the thing that most closely resembled a spirit."[58] Rousselot's version: "Speech is a movement: it is air expelled from the mouth or nose with a vibration imparted to it by the sound-producing organs."[59] A resolutely material definition. Rousselot knows, of course, that words are signs used in communication. But the interests of communication lead us to leap over the physical events of language.

In general we are concerned with *what* is said, not *how* it is said. As soon as the meaning appears clearly to the mind, we neglect the sound. . . . We have full awareness only of those sounds that correspond to a nuance of thought; we do not even notice the modifications operated on them by juxtaposition or expression. . . . When listening, we do not analyze, but are satisfied with a general impression, and if ever we have to report on the sound as heard, we supplement it by drawing from our memory the sounds that we would have wanted to produce.[60]

Communication directs perception to ends and away from means. It thus deadens us to the reality of sound, just as the alphabet and the standard national language impose their normative categories on the language as heard. Patois, because it obeys a different set of standards (which can be changed merely by traveling a few miles up the road), subverts these norms from the bottom up. The dialectologist's predicament is that the standards that can be elicited from the material (the precise *l′* of *"lapin"* as pronounced by Pierre, born three miles north of Cellefrouin in 1859) hold for only a closely circumscribed set of objects. To transcribe dialect into the national alphabet is surely to misrepresent it, but to demand that transcription realize for every dialect the ideal of "representing sounds in a fixed and invariable manner" can only lead to a proliferation of incompatible systems.[61] The Swiss dialect scholar Louis Gauchat echoed Rousselot's disenchantment with transcription:

> Our transcription signs have nothing more than a relative value: we note *ä* there where we think are hearing a more open sound than the sound we have previously chosen as the norm of the open *e*. But what is this norm? The best way to mark reference-points in the vowel scale is to take as a basis some standard words from our French pronunciation. For example, I transcribe with *ä* every dialect *e* that seems to me more open than the *ê* of the words *fenêtre* as I pronounce it. But am I always sure of my *ê*? And what if I encounter in the next village a noticeably more open *e*: shouldn't I restrict the sign *ä* to this new sound and render the previous one as an *ę*? . . . So the distribution of the signs *ę* and *ä* over the great (theoretically, *infinite!*) number of varieties of the open *e* will always be arbitrary, even in the hands of the world's best dialectologist. . . .
>
> When will the day come that the philologist carries with him a little flute that emits the standard *ę* and *ę*, when will he measure in millimeters the angle between upper and lower jaw, or the distance between the back of the tongue and the soft palate during the articulation of the sounds he seeks to study? When will we have a dial that will indicate automatically the degree of nasality of a vowel, and the like?[62]

There was certainly great variance among the speakers of so-called standard French, but the merit of an official written language is to let that be ignored, most of the time. Once the philologist ventures into the terrain of dialect, "a form of variability that is not overdetermined by the written norm,"[63] there are infinite distinctions to be drawn, each distinction holding only for an area or a time: as the alphabet reformers of the 1840s wearily conceded, there are as many vowel systems as there are speaking individuals in existence.[64] "What did I accomplish at Charmey?" asked Louis

Gauchat. "I studied, in cursory fashion, about fifty individual languages and I found *nothing individual* in them."[65] Rousselot's machines did not solve this problem so much as map it with a specificity that echoed the proliferating diversity of its objects. The results of measurement were always prone to the inaccuracies native to analog devices: loose or worn rubber tubes, laryngeal sensors strapped on too tightly, excessive volume causing the needles to skip, variations in atmospheric pressure and the like.[66] Such distortions were inevitable, given the nature of this "mode of expression and means of research."[67]

Alphabetic writing, however adorned with diacriticals and explanatory notes, now appears as a container inadequate to hold the burgeoning multiplicity that dialect scholars had discovered in orality. If the instrument set pioneered by Marey, with its novel form of writing tied directly to the physical movements of the vocal apparatus, was to be taken as the pattern of a scientific investigation of language, it must have seemed that the alphabet, with its few, crude and inconsistent notational units, held linguistics back with the weight of an ancient superstition. But as the dialect scholars admitted, the endless specification of differences, whether in villages or laboratories, was a bad infinity, condemning linguistics to mere and insufficient empirical description. The very plethora of data imposed selectivity, and that required coherent principles of selection: to which Saussure, among others, tried to respond. The solution to the problem of the missing epistemology of linguistics would begin by distinguishing levels, and first off by separating "(1) what is social from what is individual, and (2) what is essential from what is accessory and more or less accidental."[68] It would be "chimerical" for linguists to pursue the study of *parole* otherwise than as an ethnographic curiosity.[69] For "the essence of language is foreign to the phonic character of the linguistic sign."

What matters in the sign is not the sound as such, but the phonic differences whereby we can distinguish this word from all the others, for these are the bearers of meaning. . . . Further, it is impossible that sound, a material element, should belong by itself to language. For language, sound is only a secondary thing, a material on which language operates. All conventional values show this characteristic of not being identical with the tangible element that provides their support. . . . In its essence, [the linguistic signifier] is not in the least phonic, it is incorporeal, constituted, not by its material substance but solely by the differences separating its acoustic image from all the others. . . . The means of production of the linguistic sign is completely indifferent, because it does not affect the system.[70]

The expressive units of *langue* are "oppositive, relative and negative entities," defined "not by [their] material substance, but solely by the differences that separate [each of them] from all the others" (ibid.). The principle that difference of elements within a system is the only true constituent of identity explains, for Saussure, the wide leeway that speakers have in realizing particular elements in their speech—that zone of "latitude" being, more or less, the region within which Rousselot and other experimental phoneticians were marking ever finer but nonfunctional differences.

What is a relevant difference? For the experimenter, it might be a difference that can be registered by instruments; for the linguistic geographer, it might be a difference that distinguishes two dialect areas; but for a structural-functionalist, the differences that count are those that make distinctions in meaning within a single linguistic system (established as "single" for the purposes of analysis). Nicholas Troubetzkoy formulated the difference between phonetics and phonology for incipient "structuralism": "Phonetics can thus be defined as the science of the material side of the sounds of human language. . . . [But] the phonologist must exclusively consider those aspects of sound that fulfill a determinate function in language (*langue*)."[71] Bloomfield has harsh words for a descriptive phonetics without the systemic perspective:

> Practical phoneticians sometimes acquire great virtuosity in discriminating and reproducing all manner of strange sounds. In this, to be sure, there lies some danger for linguistic work. . . . The phonetician's equipment is personal and accidental; he hears those acoustic features which are discriminated in the languages he has observed. Even his most "exact" record is bound to ignore innumerable non-distinctive features of sound . . .
>
> The extent of observation is haphazard, its accuracy doubtful, and the terms in which it is reported are vague. Practical phonetics is a skill, for the student of languages often a very useful skill, but it has little scientific value.[72]

Saussure's resolving stroke was to see that the idealities of language were loose standards of performance, not entities that had to recur precisely. Neither thought nor sound could ever be adequately specified, but from the overlaid boundaries of patches cut from these two unlike kinds of material the linguist would derive the functional units necessary for analysis. "There is neither materialization of thought, nor spiritualization of sounds, but the somehow mysterious outcome is that . . . language elaborates its units by constituting itself between two amorphous masses."[73]

From this perspective, Rousselot's research looks like mere empiricism, measurement and fact-collecting. Phonetics and linguistics are not, for us, the same thing; phonetics is at most a descriptive subdivision of linguistic science; but in Saussure's assertion of the subordinate status of the investigation into the "material side" of language we can see the trace of a past threat to the integrity of linguistics. Indeed, we too easily forget that Saussure's chief theoretical contributions are responses to the problems posed by dialect and mechanical inscription. What are the "identities, realities, values" of language? If phonetics and the study of patois had dissolved the "letter" into potentially infinite subdivisions, alterations and exceptions, the new discipline of phonology would restore it in the form of an imaginary aggregation of units having only systematic mutual relations and no gaps or redundancies: the phonological system.[74] Phonology represents the return of the alphabet, restored on an ideal basis.

But Rousselot was doing something different, and subordinate only within the hierarchy of linguistic subfields established by Saussure. His investigations give an account of the "linguistics of speaking (*parole*)" left undone in Saussure's *Course*.[75] Material, analog, continuous, event-based, causal, idiographic, physiological, atomistic, this collection of traces refractory to alphabetization showed what happened when orality got a writing of its own.

The Patois of Parnassus

The Abbé was M. Rousselot who had made a machine for measuring the duration of verbal components. A quill or tube held in the nostril, a less shaved quill or other tube in the mouth, and your consonants signed as you spoke them. *They return, One and by one, With fear, As half awakened,* each letter with a double registration of quavering.[76]

The inscribing machines of the physiology lab transmit to paper a geometrical correlate of movements, or most often, of rhythms, repeated patterns of events that always return to their starting point and begin again: a cycle of four hooves touching the ground, a beating of wings, opening and closing of the mouth, diastole and systole. Each cycle or instance of the rhythm has its shape that can be compared with others; higher-order alternations and rhythms may appear. The graphic method enables us to see both the fine detail and the larger pattern.

So it was the called-for method when a controversy arose over the question whether a certain group of movements had a rhythm at all, was regular or chaotic. Mallarmé put it with mock solemnity: after tearing down and rebuilding the political order, the family, the economy, now "they've been fiddling with verse" (*on a touché au vers*).[77] Practitioners of *vers libre* passed for a band of anarchists at least as dangerous as the phoneticians, who were suspected of wanting to reform orthography.[78] The two potential sources of disorder stemmed from a single source, because French versification since its systematization in the seventeenth century was the child of the standardized written language. The dominant meter, the alexandrine, consisted of twelve syllables with an obligatory break after the sixth syllable, but what counted as a "syllable" could be expressed only on the basis of a satisfactory acquaintance with spelling. Was *lion* one syllable or two? Prose might make it one, but in verse it was always two. Verlaine's "violons" could emit "*des sanglots longs*" more readily in verse (*vi-o-lons*) than in spoken language (*vyo-lō*). Which words contained a silent *e*, and which of these silent *e*'s were obligatorily sounded as a syllable and which were elided with a following vowel, came down to awareness of spelling conventions too, as did the avoidance of hiatus (two adjacent vowels without even a mute consonant to separate them). The field of permissible rhymes was defined by spelling, grammar, and etymology (but less strictly for song and popular verse).

For the poet of 1900, the experience of the previous century had been a gradual loosening of the rules. Victor Hugo's liberties with caesura and enjambment made the alexandrine scan unpredictably: not always 6 + 6, but often 4 + 8, even 4 + 4 + 4.[79] Against such liberties Baudelaire rebelled (in demi-reactionary fashion) by heightening the classicism of his verse, leaving intact the prosody of Racine while opening up the field of subject matter to include topics obscene, debased, grotesque, or disgusting. Verlaine, with his preference for odd-numbered syllable counts (lines of nine or eleven syllables), defied the reader to place the accent unerringly, causing the poem to seem to float in time and requiring the voice to choose word-groupings and emphases differently for each line. Baudelaire had invented the poem in prose; two of the *Illuminations* of Arthur Rimbaud, "Marine" and "Mouvement," are either prose poems with poetic lineation or *vers libre*.[80] In 1886, both Gustave Kahn and Edouard Dujardin took to breaking their verse lines without obvious regard for meter or rhyme.[81] By 1900, the practice was familiar, but still suspect.

Critics complained that *vers libre* was indeterminate, neither verse nor prose, and that as a consequence they did not know how to voice it. Should the silent *e*'s be sounded or dropped? Were there caesuras within the line? Where did the emphasis go? A line by Paul Castiaux was cited as particularly frustrating, because the only way to give it a verselike regularity was to treat it like prose:

Douce et fluide caresse de cendre bleue.

If the silent *e*'s were pronounced, the line had twelve syllables, but the caesura fell in a forbidden place, between the first and second syllables of "*caresse*." Without the silent *e*'s (in normalizing orthography: "*Douc' et fluid' caress' de cendre bleue*") it formed a regular ten-syllable line with a caesura in the expected place, after the fourth syllable; but it was not obvious which scansion was correct. Confronted with this uncertainty, the critic Michel Arnauld complained that he "would prefer it if custom, in this matter of *vers libre*, told me which way to go."[82]

Robert de Souza, speaking on behalf of all writers in free verse, retorted that this only proved that Michel Arnauld was "unable to read." Moreover, "poets do not know how to read *themselves*."[83] How was it possible for poets to write what they could not read? Orthography and the metrics that grew out of it blinded them to the fact, indisputable for de Souza, that the true measure of French verse was not in syllable counts, but in accent groups, and the accent group was recoverable by paying attention to curves of intensity and duration within the sentence. Rousselot's "analytic registering and inscribing device for the slightest traces of breath and sound" offered, in this doubtful matter, a "means of control by automatic agents independent of our personal suggestions" that would lead us to "recognize the reality of harmonic and rhythmic phenomena."[84] The laboratory of "new senses" equipped by Marey's graphic method would teach poets the lost art of "reading themselves."

De Souza wrote out the troublesome verse and handed it to Rousselot without telling him his aim.

> The moment he had my sheet in his hands and had put on his glasses, he exclaimed: "What's this? It doesn't mean anything. You call that a verse? . . . I couldn't ever read that."
>
> "All right, *mon cher maître*, that's perfect! You're excellently prepared for the experiment. Hmm? Just read it out, that's all."
>
> When I turned back to check the inscription, [Rousselot] warned me: "I just read it any old way. I didn't even know what I was saying. It rubbed me the wrong way, I just didn't like it. And the mouthpiece wasn't adjusted right."
>
> He insisted on measuring himself the results of his two readings and the two other readings by M. Lote that had been more carefully performed. . . .

And what do we see? An equivalence, often absolute (or within a hundredth of a second), among the four realizations, Rousselot's two spontaneous ones and Lote's two deliberate ones. Next, a sequence of alternating long and short accents with a considerable ratio among them, more than is necessary to make them perceptible. . . . Finally, the same syllables are marked with rhythmic emotional accent on top of the regular series of iambs, emphasized by temporal lengthening.[85]

Free verse, the machinery showed, was verse: it had a repeatable, well-defined shape in time, recognized in practice even by readers who denied its existence. "De Souza reached the conclusion that a free-verse form was possible that would keep all the essential characteristics of classical verse, while delivering it from the secondary trammels to which habit had given the appearance of necessity."[86] This time, however, in a break with all previous poetic revolutions, the task of determining those "essential characteristics" was in part handed over to machines. Experimental poetry, in a new sense of the word, is what emerged from Rousselot's apparatus. Poetry was a dialect. The experiment lay in subjecting the rules of prosody to the confirmation of informants who gave evidence of this singular dialect by reading it into an "inscribing ear."

Phonetics could take the opposite tack in legitimating free verse, that is, showing that the traditional definitions of strict verse were hollow. The *Dictionary of the French Academy* (1877) tells us what an alexandrine is: "A French verse of twelve syllables when the rhyme is masculine, and of thirteen when the rhyme is feminine. Tragedies and epic poems are commonly written in alexandrines. The caesura or rest of the alexandrine verse must be immediately after the sixth syllable."[87] The words *doit être* ("must be") signal that this definition is not merely a description, but an instance of rule giving, a normative utterance. (Something of a national self-image resided in the regularity, distinct segmentation, and symmetry of the verse form.) Apart from its strictness on the matter of the caesura, the Academy's is a minimal definition, retaining only syllable count as its criterion. Other theorists of verse proposed more stringent definitions aiming at excluding flat, shapeless, prosaic, or accidental pseudo-alexandrines. The basis of French versification was the grouping of accents for some, alternation of heavy and light syllables according to others, and for others still, patterns derived from Greek and Latin quantitative meters. Furthermore, the Romantics had weakened the caesura and varied the cadence. Lines that, in classicist mode, inevitably rose to their sixth syllable and fell toward the rhyme word in their symmetrical second half (6–6) now vied with lines that had to be articulated into three or four subunits. What, then, was a good alexandrine? How did the good poets compose them? How did individual poets' practice vary? Georges Lote, a student of Rousselot, applied the graphic method to clarify these questions along with another that could not be kept separate from them: When readers performed poetry by reading it out loud, what pauses and accents did they import to the bare notation of the verse?

Lote harnessed to Rousselot's devices seventeen educated men (actors, teachers, administrators, novelists, doctors, composers, businessmen, phoneticians) and one woman who had not gone beyond primary school. (Her profession is given as housemaid—his own perhaps?) He had his informants recite into the speaking tube isolated verses and longer passages from Corneille, Racine, Boileau, Molière, and Hugo.[88] Lote's collection of tracings grew to include thousands of parallel verse recitations, measured to the hundredth of a second for syllabic duration, pauses, pitch, and changes in volume measured at mouth and nose. The result appeared in 1913 as *L'Alexandrin d'après la phonétique experimentale* (*The Alexandrine According to Experimental Phonetics*, 2nd ed., 1919) in three thick volumes, the last devoted entirely to numerical tables. Even with the limits of his selection of both verse and speakers (these latter chosen "from the bourgeois milieux . . . which make up the public of theaters . . . who still read the classics and go to see Victor Hugo's plays")[89], the richness of the corpus was unprecedented. "12 = 6 + 6," said traditional versification. Lote's equations were vastly more fine-grained. With a magnifying glass and a ruler he was able to translate the continuous scribble of the pneumatic pen tip into numbers, and then from these numbers to discover tendencies in the character of the French spoken alexandrine.

The behavior of readers led to the conclusion that the rhythm of French verse was chiefly one of duration, with the emphasized, and thereby lengthened, syllables alternating with light, unemphasized syllables. Running sometimes in tandem with this basic rhythm, sometimes against it, were a pattern of word-groupings marked by final stress and a pattern of rising and falling tone.[90] But "duration creates rhythm and is sufficient to mark it"; it is "a regulative principle dominating emphatic manifestations such as pitch and intensity . . . essentially stable."[91] Not even the most histrionic delivery could deform the "intelligent and reasonable harmony" secured by duration and "guarded by the vowels owing to their position in the temporal order."[92]

The rules of French verse as they appeared in grammars and dictionaries were not borne out by Lote's results. "Even in so-called isometric compositions, there is no constant count of the same numbers [of syllables], and the reason has to do with the normal evolution of language."[93]

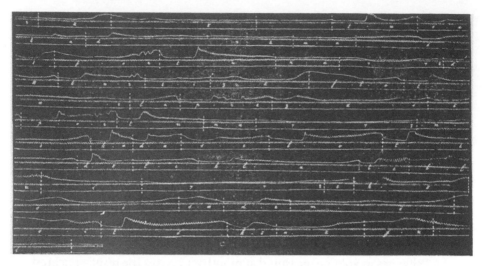

FIGURE 12. Tracings of a recitation of four verses from Corneille's *Cinna*. In Lote, *L'alexandrin*, 1: 21.

The silent *e* dwindles away; vowels formerly kept distinct in the body of a word merge into diphthongs; 12 is not always and merely 12. "The traditional alexandrine is no more than a superstition."[94] "The numerical regularity of verse is a sham. To be sure, on paper and in the intention of poets, alexandrines always have twelve or thirteen syllables. But in contemporary declamation, that number may never be attained or it may be surpassed; everything depends on the way the voice treats internal hiatus and silent *e*."[95] Prescriptive metrics, then, "either arises from an acoustic illusion or applies to a former state of the language and so has no further justification." Flexibility and a broad tolerance for exceptions: these are what the facts on the ground indicate. But if so-called traditional verse does not obey its own rules, or can do so only at the price of an artificial archaism, "free verse, far from being, as has been claimed, a monstrosity, remains viable" as an extension of the same tendencies that gave traditional verse its consistency.[96]

Lote's fastidiously detailed investigation, patterned on Rousselot's dialect studies, is like them an attempt to give a hitherto unrecorded form of orality its own writing. The oral contribution here is the attention to performance by the voice of poetic texts previously made available in print: despite all appearances, poetry is an art that requires execution, it is not complete on the page, and yet those who studied poetry had almost completely neglected its performance.[97] A tiny sampling (and by no means "representative," by our standards or those of dialectology) from the limitless body of renditions, Lote's primary material is also a form of what Saussure called *parole*, a mass of individual realizations made at their specific times and amounting to a general pattern only when overlaid one on the other by the phonetician. But Lote was not conscious of putting his thousands of examples of recitative *parole* in relation to a *langue* or grammar of poetry. No such project would have occurred to him. He was, rather, using empirical data to dislodge conventional prescriptive verse theory, an authoritative but poorly formulated body of rules. This amounted to overturning an inadequate form of writing with a more adequate one. Experimental phonetics with its curves and tracings showing the precise correlations of timing, emphasis, and pitch could capture differences that alphabetic transcription was too clumsy to record, and, according to Lote, give a more truthful account of French metrics than any that the prescriptivists had offered.

The specific form that Rousselot's and Lote's truthfulness took explains their near-absence from subsequent linguistics and semiotics. Their notations are analog, hypersensitive, idiomatic, once-off, prolix, and attain generality only with great difficulty. These shortcomings are signs of the materiality, the positivistic zeal for weighing and measuring, that the postwar generation would deem symptoms of a "naively realist deviation" and seek to replace with a definition of science that would "start from a whole and conceive of the separate elements as members of this whole."[98] The "whole" from which structuralists (led by phonology) claimed to start was simply not identical to the implicit "whole" of the phoneticians, which was physiological. The apparatus that transcribed the movements of oral speech sprang from the physiology lab. Apparatus never comes alone; it always brings concepts with it. These concepts made oral literature a product of the nerves, muscles, and sensorium, traceable and measurable to the extent that other bodily processes were.[99]

A Difference of Fifteen Cycles

> J'écrivais des silences, des nuits, je notais l'inexprimable. Je fixais des vertiges.
>
> RIMBAUD, *UNE SAISON EN ENFER*

What is a vowel? Hermann Helmholtz analyzed vowel sounds in physical-acoustic terms as

> tones produced by membranous tongues (the vocal cords), with a resonance chamber (the mouth) capable of altering in length, width,

and pitch of resonance, and hence capable also of reinforcing at different times different partials of the compound tone to which it is applied . . . . These cords . . . produce a series of decidedly discontinuous and sharply separate pulses of air, which, on being represented as a sum of simple vibrations, must consist of a very large number of them, and hence be received by the ear as a very long series of partials belonging to a compound musical tone. . . .

On investigating the compound tones of the human voice by means of resonators, we find pretty uniformly that the first six to eight partials are clearly perceptible, but with very different degrees of force according to the different forms of the cavity of the mouth. . . . Under these circumstances the investigation of the resonance of the cavity of the mouth is of great importance.[100]

Rousselot adopted Helmholtz's method of characterizing vowels by their dominant pitch and added to his laboratory a set of minutely adjustable tuning forks invented by his occasional collaborator, the Paris instrument maker Rudolph Koenig.[101] On one expedition to the field, near his birthplace in Charente, Rousselot was collecting dialect information

at Ruffec with thirty-odd people, all locals, with one exception, a vicar from the Pyrenees region. The tuning-fork had been adjusted to the frequency of the experimenter's own "closed *a*." Held before the mouths of the company, one by one, it rang out with an intense sound that could be recognized as a closed *a*. But before the Pyrenean mouth it gave out a faint and brittle sound. To get a good result, the tuning fork had to be sharpened by fifteen or so vibrations.[102]

When pronouncing the letter *a*, it appears, the vicar's "Pyrenean mouth" was habitually held at a slightly greater degree of muscular tension, resulting in a slightly higher characteristic frequency. How slightly? If the *a* pronounced by Rousselot (and the native mouths of Ruffec, ringing in unison with his) sounded at 912 hertz, the discrepancy of the vicar's *a* amounted to roughly a factor of 1 in 61.[103] But the story is not just a vindication of the tuning fork or even of the subtlety of Rousselot's experimental methods, able to detect the outsider by a barely noticeable shibboleth. It speaks, rather, of the sensitivity of the human subjects who have learned how to make the air vibrate at precisely so and so many cycles per second within a minute tolerance that expresses their affiliation to a group and a region (i.e., to a "heredity," a word Rousselot uses in a non-Mendelian way).[104] Speakers of a dialect are instruments tuned to a particular set of frequencies.

A national language is tuned to a standard, the language of the court or the capital. It centralizes and simplifies. Here dialectology has something to offer—complexity. Patois is a "tapestry," in Gaston Paris's conciliating simile, with many centers emitting their signals over small, overlapping territories. Dialectology envisions the landscape as a continuous wave-propagating surface.[105] (Gilliéron: "In the lexical world, the slightest vibration cannot arise without having repercussions on the medium where it arose, and if the vibrating element has ceased vibrating, the waves set in motion by it remain as witnesses to its existence.")[106] Rousselot ends his survey of dialect with a double verdict on speech: whatever the causes of phonetic change, "it is nonetheless certain that it is through a process independent from our will and at least in principle entirely subject to the conditions of our physical being that the sounds of language—the most precious gift of the Creator to mankind—transform themselves with admirable coherence."[107] It may have been that, seeing no way to reconcile the idioms of the priest and the professor, Rousselot simply let them subsist side by side. The Linguistic Society of Paris (which had elected Rousselot to its presidency in 1894) inscribed in its statutes, as is well known, a refusal to hear any presentation on the origin of language. The theological parenthesis ("the most precious gift of the Creator to mankind") stands outside of linguistics as Rousselot too understood it. Language might have a supernatural origin, but its natural history as here outlined goes well beyond eliminating theology from its perspective. Language is not only "subject to the conditions of our physical being," but "independent from our will" and mind. Experimental phonetics takes physical reductionism as far as it can go in its domain. "Speech is a movement"; and what science can record and seek to explain is matter in movement.

And so it does for such transient phenomena as vowels in patois. Seen materially, a language is a set of vibrations. Tuning forks may reveal differences in pronunciation, but experimental setups aside, speakers themselves are tuning forks: they emit, receive, and calibrate one another's speech vibrations. They do this not consciously, but through "a process independent of our will," one of imitation and propagation not unlike the "sympathetic resonance" long investigated by physicists.

In making speech analogous to (or identical to) a physical interaction, Rousselot seems to leap over the difference between matter and mind, between the material support of the sign and its signifying intent, or at least to demonstrate a striking lack of interest in anything but the "*unconscious* evolution" of speech.[108] Without a preliminary sense of language as a double (physical and semantic) entity, how indeed is speech possible? Or, to start from a technical prerequisite, "by what mediation does the division [of the flow of speech] into words and syllables, [a flow] given primordially to the ear as a continuum of sound, take place?"[109] Henri Bergson, exploring the relations of body and mind, admits that

> The difficulty would be insurmountable, if all that we had to go on were really auditory impressions on the one hand and auditory memories on the other. It would not be such [an insurmountable obstacle] if the auditory impressions gave organization to nascent movements that had the capacity to parse the phrase just heard and to delimit its main articulations. These automatic movements . . . vague and clumsy at first, would grow more and more defined with repetition; in the end they would outline a simplified figure wherein the listener would recognize the main features and directions of the speaker's own movements. Thus in our mind would play out, in the form of nascent muscular sensations, what we will call the *motor scheme* of heard speech.[110]

From an impossible task, that of identifying auditory impressions among a welter of new sounds differing from one another in countless ways, Bergson directs us to an achievable one, a body-to-body resonance coupling. The ear takes in vibrations and translates these into "nascent movements" of the body that would, were the hearer acting as speaker, produce similar vibrations. Understanding is imitation—bodily imitation. The hearer is "automatically" prompted to imitate the bodily gestures performed by the speaker, and dances out a "simplified figure" of the speaker's movements. The identity of the phoneme or the word is concretized by a "motor scheme" that resides in the hearer's body, a sedimented awareness of past analogous movements.

The differences from the more familiar account of linguistic identity offered by Saussure (posthumously, in 1916) are patent. For Saussure, the identity of a phoneme, word, meaning, or construction can never reduce to a single formula, however accurate or generalizable. "Their most exact quality," he says of the phoneme, "is to be what the others are not."[111] If the "oppositional, differential" account of the elements of language is the core of structuralism, then the "motor scheme" can stand for the core of the epistemology shared by Rousselot, Bergson, and other practitioners of a physiological science of culture.

One student of William James and the abbé Rousselot took Bergson at his word in making the "nascent muscular sensations" and the "motor scheme of heard speech" the solution for the central enigmas of language (how is communication possible? How is language learned and transmitted?). According to Raymond Herbert Stetson,

The normal person controls the movements of speech by the sounds produced and he perceives speech by the ear. It is probable however that the movements are the primary thing, and that what we hear is not a mere acoustic pattern but a series of movements which the sounds make audible.[112]

What we hear and respond to in each syllable would then not be the sounds, mere byproducts according to Stetson, but the muscular cycles of (citing Rousselot) *"tension, tenue, détente."*[113] Speech is, once more, a movement. For the Bergson of *Matière et mémoire*, the body and brain are the agents of memory, but not quite the places where memories are stored. That function belongs to movement; the body stores schematic patterns of reiterated movements (e.g., the precise muscle tension required to make the closed *a* of Ruffec) in its motor system.

> As we said, the body, interposed between those objects acting on it and those on which it exerts influence, is no more than a conductor, disposed to collect movements, and to send them on, when it does not halt them, to certain motor mechanisms. . . . It must therefore be as if an independent facility of memory gathered up images along the course of time as they are produced, and as if our body, with its environment, was only one of these images: the ultimate image. . . . So it is in the form of motor apparatus, and of motor apparatus alone, that [the body] can store up past actions.[114]
>
> > The brain must then be nothing more than a kind of central telephone switchboard: its task is to 'make the connection' or to postpone it. It adds nothing to what comes into it. . . . In other terms, the brain appears to us to be an analytic instrument as regards the reception of movement, and a selective instrument as regards the execution of movement. But in either case, its role is limited to transmitting and distributing movement. . . .
>
> This is to say that the nervous system is no way like an apparatus designed to make or even to prepare representations.[115]

The downgrading of "representation" and the brain, to the benefit of motor action and the peripheral nervous system, are part of the same reaction against the claims of metaphysics to direct the work of psychology that resulted in the modernist downgrading of vision chronicled by Martin Jay.[116] The replacement of thought by action, and the definition of action as muscular movement, permitting the back-translation of culture into physical motions, is a persistent theme in the late nineteenth-century sciences of man, from psychology to linguistics to sociology. Nietzsche claimed an affiliation with this current in his repeated calls for ethics, logic, aesthetics, and the like to be replaced with a "physiology" of the relevant domains.[117] Théodule Ribot, the initiator of scientific psychology in France, declares that the investigator, "seeking a leading thread to guide him in the understanding of the nervous system, learns from physiology that the reflex is the single archetype and explanation of nervous action." Well informed about German and British brain science, he quotes Jackson, Huxley, Maudsley, and others who see in consciousness nothing more than "a simple accompaniment, analogous to the shadow which follows after the body."[118] That body is not an anatomical unit but a physiological one, the body in action. If we consider "psychic facts in their concreteness" we will see "motor phenomena not as foreign elements, but as a moment of the psychic process. . . . At the root of mental life, *always and everywhere, there are movements.* Complex though the nervous organization may be . . . its basic mechanism does not change: it consists, from start to finish, in the transmission of movement."[119] The brain does not determine what the muscles will do, but is often in the position of receiving impulses dictated by them—"giving the connection," as Bergson put it, not initiating it. William James put it sharply: "we feel sorry because we cry, angry because we strike, afraid because we tremble, and not that we cry, strike, or tremble, because we are sorry, angry, or fearful, as the case may be."[120] Charles Richet insists on the reverse flow from movement to thought:

> Every movement, be it voluntary, reflex, or communicated, resounds upon the nervous centers and modifies the course of our ideas and sentiments. . . . By reason of the complexity of the cerebral and medullary acts that cause our muscles to move, we execute movements that are not always intended or conscious; some of these are reflexes; others, without being strictly speaking reflexes, are performed in a mechanical way. Consciousness of these movements does not reside in the will that commanded them, but in the muscular sensation that transmits them to the centers. . . . The muscular sense is the path whereby a great number of unconscious phenomena become conscious.[121]

In Ribot's last work (1914) the place of movement is supreme.

> My only aim [in writing *Movements and the Life of the Unconscious*] has been to call attention to the preponderant role of motor elements in the unconscious activity of the mind. . . .

Movement is in everything, everywhere, and can be the base of everything. . . .

The hypothesis I propose seems to me to result from the . . . recent studies known, since Freud, under the name of "Psycho-analysis." My hypothesis is that the ground, the innermost nature of the unconscious cannot be derived from consciousness (which is incapable of explaining it), but must be sought in motor activity, whether active or latent.

In the hypothesis that we here present, every state of consciousness is a complex of which the kinesthetic elements form the resistant, stable portion. To use a figure of speech, they are the *skeleton*. . . .

Inner speech is necessarily constituted by vocal images, reducible exclusively to the motor images of the speech organs. . . .

Memory supposes not only a modification of nervous elements, but also and especially the formation among them of associations that are determined for each particular event, the building up of certain associations that, through repetition, become as stable as the primitive anatomical connections.[122]

The translation of language, consciousness and memory into movements emulated, in the realm of theory, the laboratory apparatus of Helmholtz, DuBois-Reymond, Wundt, Marey, Rousselot, and a number of other experimentalists who set inscribing tips vibrating in response to psychological events. Reaction time, sensory discrimination, and other features of mental life could be measured in setups combining a human subject and a variety of inscribing and calibrating devices.[123] Rhythm particularly lent itself to such investigations. A strictly quantitative stimulus could be established on the machine side of the setup—for example, a series of mechanically generated clicks—and the human responses evaluated for subjective qualities such as inferred groupings or imagined accentuation.[124] At the same time as they allowed a precise measure of the capabilities of the body, laboratory devices stood in an antagonistic relation to the mind, challenging its independence from matter. Responses from the side of cognition or aesthetic value were not long in coming. Woodworth notes that

a uniformly spaced series of equal sounds . . . is often heard in rhythmic form, and the same series may be heard in different rhythms. For example, a series of seven sounds may be heard either in 3/4 rhythm or in 6/8 rhythm. These differences are not contained in the stimulus, which is equivocal. . . .

The groupings are not describable in sensory or motor terms, but are non-sensory qualities.[125]

Similarly, Wallaschek contends that

the muscular sense is not directly and in itself the cause of enjoyment in music, but becomes the cause not only of enjoyment but of high mental edification when forming the basis of a cortical process which consists in arranging a certain number of sensations in time-periods, and perceiving them as whole united groups. Through this mental process the otherwise mere sensuous enjoyment rises to the higher rank of artistic value, while without it the musical performance would have to be placed on the same level with gymnastics or, as in the savage world, with beating and fighting.[126]

In other hands, the relation was less embattled. Jean-Marie Guyau pointed out the services that the newly invented phonograph could render to psychology: it "will allow us to see not only how an object receives and stores an imprint, but also how this imprint at a given time is reactivated and produces new vibrations within the object."[127]

Upon speaking into a phonograph, the vibrations of one's voice are transferred to a point that engraves lines onto a metal plate that correspond to the uttered sounds—uneven furrows, more or less deep, depending on the nature of the sounds. It is quite probable that in analogous ways, invisible lines are incessantly carved into the brain cells, which provide a channel for the nerve streams. . . . [On reviving a memory,] the cells vibrate in the same way they vibrated the first time; psychologically, these similar vibrations correspond to an emotion or a thought analogous to the forgotten emotion or thought.
    This is precisely the phenomenon that occurs when the phonograph's small copper plate, held against the point that runs through the grooves it has etched, starts to reproduce the vibrations. . . .
    The principal difference between the brain and the phonograph is that the metal plate of Edison's still rather primitive machine remains deaf to itself: there is no transition from movement to consciousness. . . . But in fact we hear it; [in us,] in fact, its vibrations become sensations and thoughts.[128]

When we hear the phonograph, we complete the circuit, the human-and-machine circuit, that had remained incomplete without the human relay. Given that the phonograph's replay has the effect of making *us* remember, "we therefore have to concede an always possible transformation of movement into thoughts."[129]

"It doesn't get any clearer than that," says Kittler of Guyau's analogy. "All questions concerning thought as thought have been abandoned, for it is now a matter of implementation and hardware."[130] But the "hardware" did not need to wait for Edison (or Charles Cros) to come along. Physiological psychology had already turned its attention away from the brain and toward the peripheral centers of movement, exactly the junctures where a machine might be coupled with the human being to perform feedback loops and inscribe traces. Within the human being, the motor nerves had the mediating role that Kittler, for the sake of his media history, attributes to "hardware."

"Hardware" is not necessarily external to the human body. A recurrent concern of Ribot's psychology is the effort to discover the harder, more permanent, more "skeletal" layers of human memory. This distinction is formulated as "Ribot's law" in 1881:

> The progressive destruction of memory follows a logical path, a law. It descends by steps from the unstable to the stable. [Memory loss] begins with recent memories that, loosely attached to the nervous centers, rarely repeated and thus more feebly associated with the others, represent the weakest degree of organization. It ends by reaching [the level of] that sensory, instinctive memory which, anchored in the organism, has become part of it or rather has become its very self, and thus represents the strongest degree of organization.[131]

Another formulation of the rule holds that "as dissolution always reverses the path of development, it follows that the more complex manifestations of the will must disappear before the simpler ones, and the simpler ones must vanish before automatism does."[132] Amnesiacs for whom the recent past is a blank are nonetheless able to speak, dress themselves, and perform other habitual actions; a stroke patient may be deprived of most learned behaviors, and yet the autonomous nervous functions of breathing and digestion are unimpaired. The relative firmness, the psychological hardware quality, of these survival functions and, after them, habitual and motor memory is among the main structuring principles of Pierre Janet's study of hysterics, *L'automatisme psychologique* (1889). A person in a state of hysteric crisis, like a hypnotized person, suffers for Janet a deficit of the higher functions of the will and consciousness, those last developed and most fragile parts of the psychological organism. What is left belongs mainly to the bone structure of the personality, Ribot's "strongest degree of organization."

Janet's patient Léonie was apt to fall into vegetative states,

> maintaining the position in which the catalepsy caught her, without the slightest trembling to show the influence of awareness or thought. The eyes, wide open, unblinking, stared in one direction. In a word, the movements of organic life, the pulse and breathing were all that subsisted, and all movements that depend on social life and express awareness were suppressed.[133]

Patients in a state of catalepsis could be posed and moved about like department-store dummies. Or they could be induced to imitate the movements of a person in front of them.

> Rather than touch the subject, let us take a position directly in front of her, in her line of sight, and make a movement ourselves rather than shift her members. Slowly, Léonie will move her arm and then her whole body into the position that we have adopted. This phenomenon has received the name of specular or mirror imitation, because the subject imitates with her left arm the movement that we make with the right arm and thus resembles our own mirror image. . . .
>
> If I speak loudly next to [Rose] while she is in the cataleptic state, she repeats my words precisely with the same intonation. This fact is known as echolalia or echoed speech. It is most curious: the subject, transformed, so to speak, into a phonograph, repeats all the sounds that strike her eardrum, without seeming to be affected in the slightest by the meaning of these words.[134]

The mirror quality of the imitation is significant, for it indicates that the subject is not performing the same *act* as the physician (as it were, taking it up in her own right), but conforming to a pattern as presented. The transfer is not of the purpose, but of the shape, exhibited by the initiator. Sometimes Janet was able to provoke a whole drama of interconnected, habitual gestures:

> I put Léonie's hands in the attitude of prayer and her face takes an ecstatic expression. . . . I see her rise from her seat and very slowly take two steps forward. At this moment, she bends her knees, but always with unusual slowness; she kneels down, leans forward, her head bent back and her eyes raised to heaven in a marvelous ecstatic posture. . . . Now she stands up without my having touched her, she lets her head sink down and puts her raised palms before her mouth, takes five or six steps ahead even more slowly than before. What can she be doing? Now she bows with deep respect, kneels once more, raises her head a

bit and parts her lips with eyes half closed. It can now be understood what she is doing, she is taking communion. Now that her communion is complete, she rises, bows again, and with bent neck returns to kneel in her former position. The whole scene, a quarter of an hour in duration, comes to an end with the end of her catalepsy.[135]

Having become phonographs, or as here cinematographs or automata, the hysterics are certainly not amputated of consciousness, since they remain suggestible, or as we might say in technological mode, bootable. Their programs are of movements, either isolated or in connected series, and reproducible. They act out, precisely, repetitively, a loop of actions registered in somatic memory. (In this they resemble the dead people, injected with the mysterious chemical "resurrectine" by the scientist Martial Canterel, who perpetually act out the critical moments of their lives, in the 1914 novel *Locus Solus* by Janet's patient Raymond Roussel; Roussel must have read and to some extent appreciated his psychiatrist's work.)[136] Their actions are fully "organized," in Ribot's sense, leaving nothing up to chance or spontaneity.

> I now know from experience that by putting the hands of this subject into a certain position, and then leaving her alone for a few minutes, I will bring about this communion scene. But still I do not direct the scene of the communion: if I wanted to make the tiniest change to it, to have her step to the left, for example, or to have her kiss a crucifix before the communion, I would not succeed. If I speak to the subject, I am not understood, and if I touch her, I simply bring the scene to an end; I am, then, a mere spectator, not a participant. It is from her own resources that the subject draws her actions and gestures, and though she performs in so determinate a way that I can predict to the second her every gesture, she acts spontaneously. At such times is revealed more fully than ever the automatism of the subject.[137]

Even if there had been some element of play-acting—as has been said of Charcot's publicly exhibited hysterics—the degree of precision Janet attributes to these repeated performances would be difficult to achieve intentionally. Janet's description of the process is deprived of the purposiveness that we would expect in any theatrical show.

> If we have before us the series of acts that Léonie performs when I put her hands together, we think that she is taking communion and we put all these acts under the systematic idea that we know as "communion"; but it is far from certain that she [in her somnambulistic state] has the idea of communion and subordinates her actions to this

general idea; it is far more probable that she has a series of [motor] images that link up to one another, nothing more.[138]

Janet's hysterics, "transformed, so to speak, into a phonograph," correspond precisely to Guyau's definition of mechanical reminiscence: throughout their replay of past events, they "remain deaf to themselves; there is no transition from movement to consciousness." They exhibit an autographic corporeality. Memory, in their case, cannot be confused with subjectivity; it is a stock of repeated movements inscribed elsewhere than in consciousness—in their Ribotian hardware.

4 / The Human Gramophone

No man putteth a piece of new cloth unto an old garment, for that which is put in to fill it up taketh from the garment, and the rent is made worse.

—MATTHEW 3:16

The daily *La Croix*, in its number of February 3, 1927, had astonishing news for its readers: Thanks to modern scholarship, we can recover "the very words of Jesus."

> The Reverend Father Marcel Jousse, S.J., gave a series of three lectures last month that will mark a turning point in the history of exegesis. . . . The specialized study of Aramaic oral style, that is, the oral style current among the Jews of Palestine at the time of Christ, demonstrates that:
> 1. There exists, in each oral and melodically recited sentence, a unity not only of thought, but also in terms of the flexible and rhythmic balancing of propositions. These oral, living propositions answer one another in groups of two or three (more or less sharply, as it may be), in imitation of a few hundred formula-types . . .
> 2. That these rhythmic schemas themselves combine in groups of two, three, or more, and collectively form "recitatives" of greater or lesser length.
> 3. That the verbal parallelism among these recitatives can be almost total, or may amount to no more than a few words or a single word.
> 4. That these didactic recitatives are often bound to one another mnemotechnically by the use of "fastener" words, each one calling to the next.
> 5. That often an initial recitative contains, as a key, the series of fastener-words that will direct the sequence of recitatives to follow, etc.

These rules derive partly from a necessary, nearly involuntary habit on the part of the man of Aramaic oral style himself, partly from his intentional efforts. They had the effect of facilitating the memorization of long series of recitatives, retained absolutely by heart and from the first hearing. . . .

To speak only of the New Testament, it is therefore infinitely probable that the speeches of Our Lord Jesus Christ, improvised in just such an Aramaic oral style, were memorized on their first hearing and thereafter transmitted word-for-word, without the slightest alteration. . . .

When it became necessary to preach to Greek-speaking peoples, the customary Aramaic recitatives were translated into Greek. The thing translated was thus not a text read out, but a word presented for hearing and listening. . . .

[Thus,] taking two divergent Greek [Gospel] passages, the Reverend Father Jousse retransposes them into Aramaic, and their differences vanish in a single Aramaic convergence of expression. . . .

The Reverend Father Jousse, then, is convinced that our present Gospel speeches, put back into the Aramaic language according to the traditional and rigorous laws of Aramaic oral style, give us the very words of Jesus.[1]

For over two millennia in the West, writing had been seen as stable and speech as fleeting (*verba volant, scripta manent*), and when historians of texts appealed to "oral tradition" or "oral transmission," it was always to point to a gap in the documentation and acknowledge uncertainty about authorship and dates. Folklorists knew orality as a domain of proliferating variants.[2] But here is Marcel Jousse in 1927 contending that it is through orality, not despite it, that we can "prove the traditional authenticity of the Gospels and Epistles, and thence, thanks to the ethnic fact of invariable memorization of oral style, arrive at their authors themselves."[3] This reversal of the positions of orality and writing calls for an explanation.

"Errores Modernistarum"

The "Modernist crisis" in Roman Catholicism, touched off in 1902 by the publication of the abbé Alfred Loisy's *L'Évangile et l'Église* (*The Gospel and the Church*), was a crisis over the status of texts.[4]

The Gospel and the Church was a rejoinder to Adolf Harnack's *Das Wesen des Christentums* (*The Essence of Christianity*, 1900). In a series of lectures to first-year university students, Harnack, aware that religious practice had long been falling off among the educated, advanced an argument designed to appeal to skeptical idealists: The Christian message consisted of a moral "kernel" and a historical "husk."[5] The kernel consisted of a few supreme doctrines: "the kingdom of God and its coming . . . God the Father and the infinite value of the human soul . . . the higher righteousness and the commandment of love."[6] In comparison with them, the Bible's genealogies, the miracles, the quibbles of the law, the location of Eden, the duration of the Flood, the oxen, asses, maidservants and sandals, not to mention councils, priests, liturgies, vestments, statues, candles and the like, were mere historical details. The kernel was eternal and essential; the husk, circumstantial and dispensable. "The fact that the whole of Jesus' message may be reduced to these two heads—God as the Father, and the human soul so ennobled that it can and does unite with him—shows us that the Gospel is in nowise a positive religion like the rest; that it contains no statutory or particularistic elements; *that it is, therefore, religion itself.*"[7]

That, for Harnack, was the true religion of Jesus, one that might be received by modern, post-Kantian people.[8] But in the centuries separating Jesus from us, the original insights have been covered over with misapprehensions, Harnack contended. The direct intuition of the Kingdom was replaced by a complex system of dogma ringed round with sharp anathemas; an obsession with eternal life, whether as reward or punishment, covered over the commandment to love one another; and an institution, the Church, arose to administer the lives of the faithful and eventually amass temporal power.[9] Yet, periodically true religion reemerged, tearing away in a "critical reduction to principles" what had become dead tradition.[10]

In response, Loisy observed that Harnack would be singularly fortunate if he could square his ambition to detect the "essence" of Christianity with his task as a historian. If it were a matter of discovering the *historical* core of the Gospel, Loisy insisted, the only sound method would have been to start from the best-authenticated, earliest texts and show the meanings they had in their original situations; lacking these, Harnack's method "might have exposed its author to the supreme misfortune, for a Protestant theologian, of having founded the essence of the Gospel on an attribute of the [post-Jesus] Christian tradition."[11] Harnack would have been on surer ground if only

Christ had composed himself an account of his doctrine and a summary of his preaching, a methodical treatise on his mission, his role, and his hopes; then the historian could give this document the most careful consideration and determine the essence of the Gospel from this unquestionable testimony. But no such document ever existed, and nothing can stand in its place. . . . Whatever we may think, theologically, of tradition, whether we cling to it or reject it, we know Christ only by the tradition, through the tradition, in the early Christian tradition.[12]

And that tradition was likely to resemble anything but a fairy tale picture of religion within the limits of reason alone. Loisy proposed tracking back to the earliest texts of the movement, insofar as they could be distinguished from later additions, and asking what the claims of Christianity had been at different stages of its development; in this way we might retrace the path that led, unpredictably, from "the Gospel" to "the Church."

Loisy claimed to have written "an essay in historical construction," not "a system of theological doctrine."[13] But by choosing to frame his defense of Catholicism as a retracing of the stages by which the Church grew out of and away from the first Christian communities, Loisy trespassed against the doctrinal prerogatives of the church in which he still exercised priestly functions. In 1893 he had already been dismissed from his teaching post at the Institut Catholique for introducing the methods of historical biblical criticism and mentioning some of their results, such as that "the Pentateuch, in its present state, cannot have been written by Moses; the first chapters of Genesis do not contain an exact and factual history of the origins of humanity; not all the books of the Old Testament have the same historical tenor."[14] Although L'Évangile et l'Église was intended as "anti-Protestant and neo-Catholic,"[15] Loisy's portrayal of Catholic tradition *as a tradition* gave the defenders of institutional inerrancy their opening. Loisy's book was condemned as heretical by the archbishop of Paris (January 1903). He had the imprudence to follow it with another short book, *Autour d'un petit livre*, that explained in greater detail the difficult points of the first. In July 1903, Leon XIII died, and his successor, Pius X, named traditionalists to important posts in the Vatican machinery. Loisy's books were analyzed by a committee of the Holy Office and put on the Index in December 1903. As he refused to make a full retraction, he was excommunicated and declared *vitandus* (a person it was necessary to avoid) in March 1908.[16] The Collège de France, the country's most prestigious secular institution of learning, was quick to name Loisy to a professorship of history of religions (1909). In July 1907, a decree of the Holy Office, "Lamentabili sane exitu," condemned sixty-five "errors of the Modernists," almost all taken (or tendentiously paraphrased) from Loisy's work. Heretical "Modernism" included the following opinions:

2. Although the Church's interpretation of the sacred books is not to be rejected out of hand, nonetheless it should be subjected to the more accurate judgment and correction of scholars.

11. Divine inspiration is not so extended to the whole of sacred Scripture that each and every part of it is protected from any error.

16. The narrations of John [the Evangelist] are not truly history, but a mystical contemplation of the Gospel; the speeches in this Gospel are theological meditations on the mystery of salvation, deprived of historical veracity.

38. The doctrine of the Atonement is not in the Gospels, but is Paul's creation.

53. The organic constitution of the Church is not immutable; rather, the Christian community is subject to perpetual evolution, just as a human community is.

64. Scientific progress requires that the Christian doctrinal conceptions of God, of creation, of revelation, of the person of the Incarnate Word, and of redemption undergo change.[17]

A subsequent papal encyclical, "Pascendi dominici gregis," restated the "errors of the Modernists" and was followed by an oath against the pernicious doctrines of Modernism, to be sworn by all clergy with teaching responsibilities.[18] To read "Pascendi," it would appear that the church was about to be enveloped and crushed by a monstrous conspiracy hatched in its own breast yet hiding under the most inoffensive appearances. "Pascendi" performs, in Pierre Colin's words, "the invention of Modernism."[19] This more than a little imaginary "modernism" concentrated in one figure the forces against which the Church saw itself as contending: rationalism, socialism, materialism, secular government, evolution, feminism— and philology.[20] The authority of religion in European society and politics had been weakening, of course, for some time. Science and scholarship had developed into self-regulating professions. The 1905 laws codifying the separation of church and state in France were passed between Loisy's first and second censures. To such challenges from outside, the Church seemed to respond only by tightening its interior discipline.

There was no place, in other words, among Catholics for the kind of study of scripture that, initiated by Spinoza and Richard Simon in the seventeenth century, had become a recognized professional specialization in nineteenth-century England and Germany.[21] What passed for "orthodoxy" was the stubborn insistence that the truth had always been manifest and identical to itself, in texts and in fact. With a fine disregard for the sensitivities of his superiors, Loisy invoked the results of textual and historical criticism as a matter of course.

Loisy's two antagonists, Harnack and the Papal Curia, different though they are, share a theory about what a text is. A text has well-defined borders and conveys a single intent from a single author. Such a text, Loisy's historical research reminds us, is fantasized and impossible in a historical world. A "text" in Loisy's understanding is a bundle of discrepant intentions made to hold together by external and internal pressures, a patchwork that a more searching reading can always tear apart and recompose. A "text" in Harnack's sense, when rightly pared of its "husk," consists of an "essence," its glorious body. And the Curia's text is and always was that glorious body, divinely inspired and inerrant. There is no essence to a text as understood by Loisy, there are only contingencies.

Proposition 15 of "Lamentabili" rebukes New Testament scholars for holding that "The Gospels were expanded by continual additions and alterations until the definitive formation of the canon; thus in them there remains of the teachings of Christ only a faint and uncertain vestige." Indeed, Loisy had said that "what remains in the Gospels is only an echo, necessarily diminished and somewhat confused, of the words of Jesus: We have the general impression he produced on his receptive hearers, and the most striking of his formulations, as they were understood and interpreted at the time; and there remains the movement that Jesus began."[22] Try as he might, Loisy could not frame a believable "orthodox" version of the textual origins of the New Testament: Was the Vatican's purpose to assert that "even before the definition and formation of the canon, the Gospels were always exempt from additions and corrections, and therefore in them the teaching of Christ has remained complete and certain"? That would be "exaggerated to the point of flirting with error," he observed drily.[23] But in the tide of pamphlets and books written to rebut Modernism, one opponent, Father Gaston Oger, found a new technological figure of speech to circumvent Loisy's "diminished echoes":

> In matters of faith, texts guaranteed by the authority of the Church cannot be considered to lack warrant, and without such authority, there can be for most texts no undisputed wording and no sufficiently clear meaning. . . . The apostles were echoes, and their teaching an echo of the Master's teaching, but an echo that altered or attenuated

not a one of the original waves. More faithful than the phonograph, they recorded not only the Master's words, but his timbre, and nothing was lost for the generations to come.[24]

The Gospel of Movement

> But we have this treasure in earthen vessels.
>
> PAUL, II CORINTHIANS 4:7 (cited on the last page of Jousse's *Style oral*)

Oger's "gramophone" simile was pure bravado, a way of stating as fact what would have to have been true for certain axioms of church doctrine to be at all probable. But twenty-two years later Jousse's *Le style oral* was published, bearing a red wrapper on which was printed the publisher's announcement:

> This stunning and powerful experimental synthesis brings us a whole new scientific method, destined to create a sensation in the scholarly world of which the most recent discoveries are here used, combined and illuminated. Among other problems touching on Thought, Language, Memory, Rhythm, etc., and resolved by the implacable logic of facts, the reader will see here reduced to nothingness, in a new and unexpected manner, the superficial hypercriticism of M. Loisy concerning the texts of the New Testament.[25]

In Jousse's hands, orality was called to new tasks. Drawing on the work of Rousselot, Janet, and Mauss, *Le style oral* innovates in proposing an idea of oral tradition that is not privative (that is, mainly characterized by the lack of writing), but positive: an attempt to model the specificity of the medium, to say what orality is and does rather than what it is not. Oral tradition, for Jousse, will not transmit just any message, but shapes the messages it conveys in particular ways that are accounted for by the material conditions of the transmission. Texts, Jousse held, were unreliable because any written document is wrongly structured as a representation of the content that oral traditions transmit.

If Jousse offered a theory of oral composition and transmission in which the lack of writing was not the major theme, it was because the scope of his ambition was large. Once the devices specific to the oral medium had been described in their own terms, literacy would appear to be an untrustworthy witness to orality, rather than the other way round. "Our Greco-Latin civilization of letters" would be dislodged by "the ethnic laboratory." The "errors of Modernism," for Jousse, were only one derivative symptom of the reliance on writing—a forgetting of everything that memory had been trained to preserve—and Jousse came to restore memory. This would require more than one transformation.

The lectures to which *La Croix* referred were given at the Pontifical Biblical Institute in Rome in January 1927. This Institute was the seat of the Pontifical Biblical Commission created in 1903 to prevent future Loisys from engaging in unauthorized philological analysis. Its pronouncements, particularly when countersigned by the Pope, had force of law in Catholic schools and seminaries. Jousse's conclusions were welcome. *La Croix* concluded its report of the event with a box score: "Even our hypocritical adversaries admit that some of their most cherished negative conclusions are unsettled today by Jousse's arguments."[26] Conspicuous in the large audience at Jousse's Rome lectures was Cardinal Louis Billot, who had been one of the main authors of the encyclical condemning Loisy in 1907.[27] Pius XI, in a personal audience, told Jousse that his *Oral Style* was "a revolution, but common sense too." Jean-Baptiste Frey, C.S.Sp., secretary of the Biblical Commission, presented the lecturer with a book inscribed: "To the Reverend Father Jousse who by a new path confirms ancient truths." Leopold Fonck, S.J., the founding director of the Institute, who apparently had some reputation as an enforcer, declared that Jousse had "proved the authenticity of the *Magnificat*." Another Roman authority commented, "If Jousse had come along fifty years earlier, Modernism would never have happened."[28]

Precisely: The aim of Jousse's theory of orality was not to attack Modernism, but to take away its conditions of possibility. If the Gospels disagree among themselves, let the Greek texts be only so many poor translations of an Aramaic original.[29] If the historicity of Jesus comes into question, let documentary history retreat before a sung and recited history still reverberating from the Lord's original impulse. If the relation of the Gospels to the later Church is doubtful, let this be the occasion of redefining the Church in the image of the restated Gospels. Thus, Modernism will never have happened. It will have been at most an error parasitic on an earlier error, the error Jousse comes to correct by restoring the oral style as the medium of the Bible. What the antimodernists in their jubilation may not have noticed is that the territory that Jousse takes away from Loisy becomes his own, to do with what he will.

The *Magnificat*, singled out by Jousse's Roman listeners as a triumph of his method, furnishes a convenient example. One of Loisy's first publications dealt with a textual problem adjoining the *Magnificat* for which the Vulgate's authority was insufficient. Several manuscripts of the Latin Bible predating Saint Jerome's editorial work, as well as citations in other early texts, introduce the song of thanksgiving and praise as follows:

> And Mary arose in those days ... and entered into the house of Zacharias, and saluted Elizabeth. And it came to pass that when Elizabeth heard the salutation of Mary, the babe leaped in her womb, and Elizabeth was filled with the Holy Ghost. ... And *Elizabeth* said, "My soul doth magnify the Lord, and my spirit hath rejoiced in God my savior. For he hath regarded the low estate of his handmaiden: for behold, from henceforth all generations shall call me blessed. For he that is mighty hath done to me great things, and holy is his name. And his mercy is on them that fear him, from generation to generation. He hath showed strength with his arm, he hath scattered the proud in the imagination of their hearts. He hath put down the mighty from their seats, and exalted them of low degree. He hath filled the hungry with good things, and the rich he hath sent empty away. He hath holpen his servant Israel, in remembrance of his mercy. As he spake to our fathers, to Abraham, and to his seed forever."
>
> And Mary abode with her about three months, and returned to her own house.[30]

Jerome's Vulgate, together with virtually all texts stemming from it including Protestant translations, assigns the speech to Mary, so one small task for philology will be to account for the discrepancy. Loisy hypothesizes that the Greek originally read καὶ εἶπεν, "and [she] said" with no explicit naming of the speaker; later copyists, ill at ease with the gap, must have decided who the speaker would be.[31] Against the by now long traditional assignment of the speech to Mary, Loisy argued that Elizabeth had equally good, if not better, reasons to speak in these terms, as she was about to bear a son (the future John the Baptist) despite her advanced age. Elizabeth's situation, more than Mary's, replicated that of Hannah in the first book of Samuel, and the wording of the *Magnificat* closely repeated that of Hannah's song of thanksgiving:

> And Hannah prayed, and said: My heart rejoiceth in the Lord, mine horn is exalted in the Lord, my mouth is enlarged over mine enemies, because I rejoice in thy salvation. ... They that were full, have hired themselves out for bread, and they that were hungry, ceased; so that the barren hath born seven, and she that hath many children, is waxed feeble. ... The Lord maketh poor, and maketh rich; he bringeth low, and lifteth up. He raiseth up the poor out of the

dust, and lifteth up the beggar from the dunghill, to set them among princes, and to make them inherit the throne of glory.[32]

"The content of [the *Magnificat*] has no personal connection to Mary," says Loisy, and indeed, apart from a sentence or two,

> the *Magnificat* is nothing but a replica of the hymn of Hannah, mother of Samuel. . . . Neither the *Magnificat* nor the *Benedictus* [uttered by Elizabeth's husband Zachary, Luke 1:68–798] were taken down stenographically at the dictation of the characters to whom they are credited. . . . Is it not more natural to suppose that the composer of the hymns would not have needed to have recourse to Hannah's psalm to express the feelings of Mary, whereas, if he wanted to bring out those of Elizabeth . . . the psalm of Hannah stood ready as a model to follow? . . . If the editor had made Mary speak, he would probably not have imitated the old model so closely and would not have needed to borrow all its ideas to fill out his framework.[33]

Textual criticism as Loisy understands it requires that we put ourselves in the place of the "composers," "editors" and "copyists," who refer back to other parts of sacred writ in order to "borrow ideas" and "fill out frameworks." In the absence of ancient Galilean "stenography," what else could anyone imagine? This pattern is exactly what oral theory dissolves. Jousse's hypothesis of collective memory and recitation turns the specific, sequential acts of inscription necessary to Loisy's reconstruction into a pool of phrases waiting to be combined in a traditional structure, of which he gives an idea in a kind of reverse cento of parallel passages:

> Attentive study and the methodical classification of facts show us that all the Rhythmic Schemata of Recitations are, for a given ethnic milieu, nothing more than "the repetition, under slightly differing forms," of "four to five hundred typical {schemata}," "the traditional and primitive elements" of the Oral Style (Jean Paulhan, *Les haintenys mérinas*). This was noticed long ago in the case of the admirable improvisation of the young Virgin Mary: "The *Magnificat* is the work of a soul deeply familiar with the sacred texts, {most probably in their Aramaic oral targum and learned by heart}. Its expressions and often its ideas are borrowed from the writings of the Old Testament . . . " (Lesêtre, cited by Vigouroux, *Dictionnaire de la Bible*) . . .

> Exultavit cor meum in Domino et exaltatum est cornu meum in Deo meo.[34]

> Benedic anima mea Domino.[35] Exultabit cor meum in salutari tuo.[36]

Laude anima mea Dominum.[37] Neque despicias me Deus salutaris meus.[38]

Magnificate Dominum mecum.[39] Et defecit spiritus meus.[40] Etc.

Magnificat anima mea dominum et exultavit spiritus meus in Deo salutari meo.[41]

The possible assignment of the speech to Elizabeth, though supported by some of the manuscript witnesses, does not even appear in Jousse's comments, as what matters for him is inventorying the collective subconscious repertoire of topoi from the midst of which Mary would have spoken (and it must be Mary who speaks, in Jousse's rendition, for she demonstrates the maternal education in Aramaic rhythmo-catechism that the young Yeshua would have received before embarking on his ministry).[42] The criteria for authorship, authenticity, and consistency, and what counts as a doublet, a gap, or an interpolation, differ once we are circulating in the domain of oral style. The tags of earlier and later, original and copy, author and nonauthor, fade away in an economy of mutually substitutable parts and traditional lines of thought. Orality stands in a different dimension of time: an oral tradition is a reservoir, a virtual coexistence, of particulars that we are accustomed to see separated into distinct, mutually exclusive events (the column of textual parallels in the passage above standing for this well of memory). Did Hannah compose the *Magnificat*? Or was it Mary? Or Luke? Or the early Christians? The answer is: Yes. As competent speakers of the relevant cultural "langue," they qualify as having potentially composed it, whenever the occasion might have demanded. Jousse's presentations are meant to bring us into that frame of mind, as if we too could have juggled those materials and made the same utterance. Once one "holds in memory or in *synoptic tables* all the traditional clichés of the Hebrew and Aramaic oral style, with their familiar plays on words, one can . . . pierce the dried husk of our poor French word-equivalents and revive in ourselves, with some of their Aramaic verbal subtleties, the gentle Balancings of the didactic Recitatives improvised and wrought in order by the Divine Oral Composer."[43]

The battlefield on which Modernists and traditionalists contended was constituted of texts. They might be judged authentic or inauthentic, and that on transcendental or historical grounds, but they were definite sequences of words that had been set down, at definite times, on parchment or paper. The oral style perturbs this field. For it, as Kittredge said, "there are texts but there is no text"; its works are characterized by collective composition, formula, iterability, and virtuality. With Marcel Jousse, those characteristics frame the most eminent text of all, the most fought-over, and make it something other than a text.

Le style oral brought its author sudden celebrity.[44] The abbé Henri Bremond, of the Académie Française, called on Jousse to clarify questions of rhythm in the ongoing debate over "pure poetry";[45] the literary critic Frédéric Lefèvre interviewed him for *Les Nouvelles littéraires* and devoted a volume to exploring his "new psychology of language"; part of Lefèvre's enthusiastic commentary was reprinted in the series *Le Roseau d'Or*, directed by Jacques Maritain, the theologian best admitted in literary circles at the time.[46] Each time, Lefèvre put extraordinary emphasis on the statement "We have the very words of Jesus": both attention-getting and a badge of Catholic orthodoxy.[47] Lefèvre even forced a thirteen-page digression about Jousse into a book of dialogues with the Catholic philosopher Maurice Blondel, to the latter's irritation.[48] A manual of "notions on language according to the works of Father Marcel Jousse" was published for the use of philosophy students.[49] The revised *Handbook of Psychology* incorporated his terminology.[50] Theologians and historians of religion acknowledged Jousse's theories, fully, strategically, or with objections. Léonce de Grandmaison, the editor of the Jesuit periodical *Études*, announced that the question of rhythm in the New Testament "has now entered a new stage, thanks to an extension which broadens the field of comparison and to a method which replaces more or less vague intuitions, chiefly literary in nature, with the study of the linguistic and psychological bases of this type of composition."[51] Alfred Loisy referred distantly in 1933 to "a dissertation with the bizarre title of *The Rhythmic and Mnemotechnic Oral Style of Verbo-Motor Individuals*, the continuation of which has been long awaited," and claimed priority in publication of its best ideas.[52] At the Théâtre des Champs-Elysées, a group of little girls, trained in the "rhythmic recitations of Rabbi Iéshoua and his apostles," performed synchronized chants and gestures before an audience of society people and churchmen.[53] From the early 1930s to the middle 1950s, Jousse presented his findings weekly before an auditorium at the Sorbonne, at the École d'Anthropologie, and in a seminar for teachers of young children.[54] Even *Time* magazine, with its customary exaggeration, sketched Jousse at work.

To the Society of Jesus, militant defenders of Roman Catholic orthodoxy, a French Jesuit named Marcel Jousse has long been its *enfant terrible*. A onetime artillery captain who began studying for the order after World War I, white-haired, fiftyish Père Jousse invented and today teaches something he calls Rhythmocatechism, or preaching with gestures. . . . When Père Jousse lectures, 200 people watch goggle-eyed: doctors, spiritualists, philologists, ballet students, poets (among them Paul Valéry)—and two Jesuit theologians, hawklike for heresies.[55]

That the claims of orality found such resonance is not to be explained by their sheer intellectual content, rather by their coincidence with certain themes in the cultural and spiritual life of the time, particularly in France. Primitivism may have contributed.[56] The "return to order" following the slaughter of the Great War took, for many, the form of a conversion not only back to their prewar faith, but to a religion more demanding, more archaic, and more literalistic, a Thomism ready to organize every aspect of the artistic, moral, and spiritual life of the age.[57] Among traditionalists in art and politics, Jacques Maritain's *Art et scolastique* (1920) and Charles Maurras's daily *L'Action française* set the terms of debate. Maritain: "It is well known that Catholicism is as much *antimodern* by its immovable attachment to tradition as it is *ultramodern* in its boldness in adapting to new conditions emerging in the life of the world."[58] Aesthetic forms previously associated with the avant-garde became vehicles for expressing disenchantment with the modern condition (with *The Waste Land* the most conspicuous Anglophone example). As Stephen Schloesser notes,

> non-Catholics, non-Christians, and non-believers played significant roles in Catholic [cultural] revivalism. The success of Bernanos's *Under Satan's Sun* was largely due to a popularized notion of 1926 as the year of the 'Catholic novel'—an idea given cultural weight by Albert Thibaudet's essay published in the thoroughly secular *Nouvelle Revue française*. Similarly, Bernanos's reputation . . . depended on the activist advocacy of Frédéric Lefèvre, editor-in-chief of *Les Nouvelles littéraires*, the most influential postwar literary review. . . . In sum, cultural critics thought through philosophical, pictorial, literary and musical productions using variations on peculiarly Catholic ideas.[59]

The theory of the oral style was perfectly designed to break through the skepticism of moderns (not to mention Modernists) toward the Jesus tradition. It changed the subject. Though the Four Gospels on paper are rife with contradictions and inconsistencies, and all the more so for those who can read them in Greek, the perspective of oral tradition might reconcile them at the level of "deep structure" as if they were merely alternative translations into language (not only the Greek language) of one and the same gestural content. In the structures of rhythmic recitation Jousse thought he had found a new channel of communication impervious to the philologists' skepticism, a means whereby the historical Jesus might still speak to us. But it was not to be in our modern language; such language was debased and unreliable, a source of "pseudo-problems";[60] rather it would be via the rhythms, as it were the talking drums or Morse code, fossilized in texts.

Our flattened, debased, bloodless modern language, even in the extreme "algebrosis" of silent letters running across paper, could nonetheless be made to dance to the rhythms of Aramaic recitation as Jousse conceived it. Typography could be made to reflect the profound degree of organization of a text made to be chanted and learned: its parsing and chunking of the message, its semantic saliencies, its pointing of antithetical or synonymous terms. Jousse's rhythmo-typography reflects his analysis of passages of Scripture into component gestures each corresponding to a bodily cue: on the *one hand*, on the *other hand*; the first *step*, the second *step*, and so forth, in a regular rhythmic alternating cycle. It would amount to an autography of the gestured Gospel. Orality once more colonizes the realm of two-dimensional inscription by introducing its codes and message. "The signifying exemplarity of this layout remains as a silent mark of the gesturality of its author"[61] (or its translator), who has "mapp[ed] the oral event onto an augmented textual surface designed to bear more and different kinds of meaning than can the conventional printed page . . . a kind of overdetermination of the reader's activity . . . transform[ing] the [imagined] reality of the oral traditional event into a visual, textual rhetoric, specifically a set of typographic cues, with the express intention of mitigating the displacement caused by textualization of that event."[62]

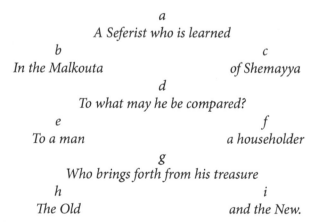

a
A Seferist who is learned
b   c
In the Malkouta   of Shemayya
d
To what may he be compared?
e   f
To a man   a householder
g
Who brings forth from his treasure
h   i
The Old   and the New.

New, [continues Jousse,] in the unexpected juxtaposition of brilliant and more than brilliant formulae. Old, in the traditional juxtaposed Formulae.[63]

The segments marked *a*, *d*, and *g* stand for a "raising" of the Burden, whereby the student takes on the weight of the chant, and the paired segments *b–c*, *e–f*, *h–i* stand for the "balancings," whereby the weight of the Burden is distributed in manageable packets. The raising of the Burden has the form of the body's forward-backward axis, and the balancing, that of the left-right symmetry of arms and legs. By parsing the entirety of Holy Writ into such "Rhythmic Schemata," Jousse thinks, we can recover the motor schemata that made it so easy for the ancients to learn their oral texts and so hard for them to forget them.

Reconstituting orality, as Jousse wants to do, implies undoing much of the history of Christian doctrine through a back-translation to a language of Judeo-Christian common origins.[64] (Like Harnack, Jousse desires to recover a Gospel before the Church.) Just as the Jews, after their return from Babylonian captivity, found themselves obliged to appoint translators or *metourgemans* in the synagogue to paraphrase the Hebrew readings into the Aramaic language understood by the congregation, so in the early Christian churches, Jousse supposes, translators were on hand to render the Aramaic rhythmical recitatives in Greek (using traditional Hebrew-Greek equivalents established by the Septuagint).[65] From the loss of contact with the original Aramaic and the spread of an imperfectly glossed Greek and Latin version of the religion dates the dominance of "pseudo-problems" such as would make the career of Loisy. Aiming to reverse this (polemically elaborated) history, Jousse becomes our *metourgeman* and reads the Gospels gesture by gesture, spreading the elements of gesture out on the page as if to choreograph their delivery and paraphrasing them in his idiosyncratic "ethnographic" style. Take, for example, Matthew 11:28–30:

> Come unto me all ye
> who are burdened   and heavy laden
> and I will give you rest.

[Jousse's commentary:] That is to say, I will give you lessons that are short and easy to retain, to memorize. You are burdened and heavy-laden because you carry a recitation-burden that is too heavy. This reading is furthermore confirmed by Jesus, who says:

> Take upon ye   And be ye
> my yoke   learners of me[66]

Jousse reads the "yoke" here as a picture-analogy of the parallel, symmetrical mnemonic verse, as well as of the learner's voluntary servitude to his teacher: to learn is to be attached to the teacher's mnemonic yoke.

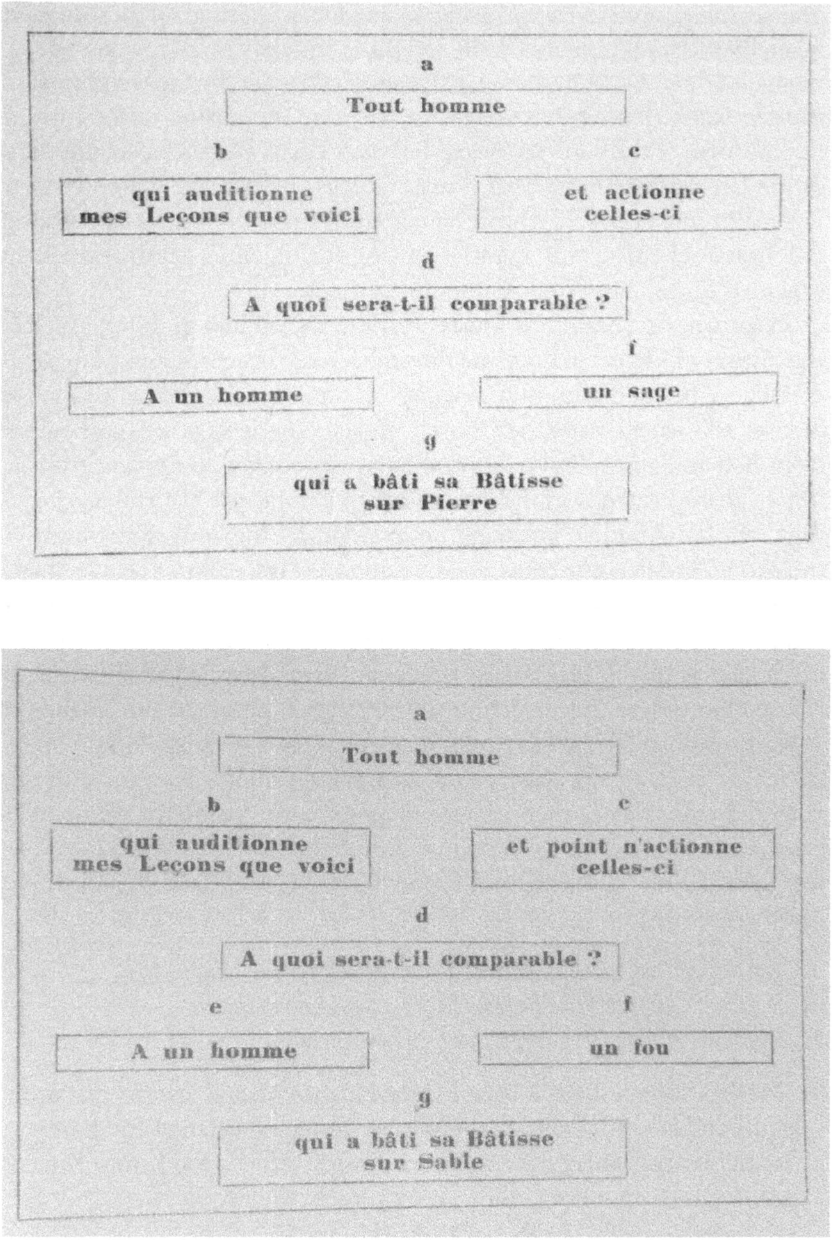

FIGURE 13. Rhythmo-typography. From Jousse, "Père, Fils et Paraclet dans le milieu ethnique palestinien," 26–27.

> *For I am simple   and brief for the memory*
> *and you will find rest for your [reciting] throats.*

[What I translate as "simple and brief for the memory," says Jousse,] you might translate as "gentle and lowly in heart." Well, there are doubtless seven and seventy meanings in any Hebrew phrase, so you can pick and choose. But there is one meaning that is fundamental, after all:

> *For my yoke is easy   and my burden is light.*

This law of the [mnemonic] yoke is the very pedagogy restored by that great renewer and regulator of gestures whom we know as Rabbi Yeshua the Galilean.[67]

This allegory of self-reference, this midrash (as Jousse proudly calls it) that sees the Gospel as a description of its own methods of teaching, reorients the messianic import of the book. "Life eternal" is achieved in memory and by passing on the yoke to future memorizers. The "soul" that one must retain even at the price of losing the world becomes, thanks to a complex etymological play on the Aramaic word *nāfshā* (Hebrew *nefesh*), "breath in the throat, life," the reciting throat, trained to reproduce automatically the rhythms of the teacher. The Jesus who is recovered is simply the voice of oral tradition itself, and the message of the Gospels is shown, through this recoding, to be the message that their oral medium was bound to convey. "A magnificent task remains to be done: show that all the terms of Yeshua in the Gospel point toward memorization."[68] In other words, the message is the medium.

> *If you do not repeat   and do not become once more*
> *   like schoolboys*
> *Never can you enter   in the Malkoutā of Shemayyā*[69]

(Malkoutā: rule, kingdom; Shemayyā: blessedness.)[70] "If you will retain the precepts of me, you will remain in the love of me, just as I have retained the precepts of the Abba and I remain in the love of the Abba. . . . Greater love hath no man than this, to give his reciting-throat for his friends."[71] As did Rabbi Yeshua. The "Abba," the Father, receives a disciplinarily specific translation in the idiom of Palestinian recitation. Any teacher, any Rabbi, says Jousse, stands in the relation of an Abba to his Berā, or Learner. And "every learner must recite in the terms of his Rabbi."[72]

There is thus a trinity of roles for this oral transmission of the Tōrāh (in Aramaic, the Orāyetā) in the father's house: the *abba*, the *berā*, and the *paraqlitā*. This technical term of *paraqlitā* or "ad-vocatus," intermediary, interpreter, is an Aramaicized Greek word. In the Targums, it was traditionally used like the Aramaic word *metourgemān* to gloss the Hebrew word *mélitz*, "interpreter."[73]

"Speech is the 'son' of the Speaker, and Breath proceeds from the two of them, like a phono-mimetic echo."[74] So much for the mystery of the Trinity. In this way, dogmas, mysteries, rewards, and punishments are transfigured by their translation into "the gestural meaning of Palestinian words."[75]

In the Torah, everything is a Gesture, everything is *Dabar*, but not everything is speech. This polysemantism comes into play, precisely, when one knows that a word is just a diminution of an action. We find this gestural and recitational mechanism sublimated in Yeshua, that great structurer of gestures, this great praxic, he who brought the Rule, the Malkouta, that is, the Rule for action which we have already cited in keeping with its own midrashic logic:

> *You will learn*   *and you will retain*
>   *and therefore you will love*
>   *The Lord*   *your Teacher*
>   *with all your memory-heart*
> *with all your reciting-throat*   *and with all your miming muscles.*[76]

The Law and the Prophets having been composed in hain-teny fashion from the beginning, by reiteration, rhythmic variation, and reorganization of prior recitals, Rabbi Yeshua and his followers were simply carrying the tradition forward. "Think not that I am come to destroy the law or the prophets," said Rabbi Yeshua: "I am not come to destroy, but to fulfill" (Matthew 5:17). Marcel Jousse, too, comes to show the so-called New Testament as the replenishment of the inherited Torah, and Rabbi Yeshua as its Homer. "It is incumbent on us to embody in ourselves the living pedagogy of Yeshua in order to understand the liberating pendulum-swings of that young Galilean peasant, of that Meshīhā who . . . brought to the world the most sublime regulation of human gestures."[77]

>   *a*
>   *Go ye therefore*
>   *b*   *c*
> *and rhythmo-catechize*   *and immerse*
>   *all the Goyim*   *them*[2]

The rhythmo-typography must be an invitation to join in the dance.

A Bone Gallery

> If one wanted to put it rather drastically, one could say that your study is located at the crossroads of magic and positivism. This spot is bewitched. Only theory could break this spell.
>
> ADORNO TO BENJAMIN, 10 NOVEMBER 1938

To reach the room where "the Professor" (as he preferred to be called) Marcel Jousse lectured on linguistic anthropology at the École d'Anthropologie in Paris one had first to walk down a long hall lined with glass-faced cases from which an intimidating collection of skulls and mounted skeletons, some of which had belonged to earlier members of the École, stared out at the passing public.[79] Jousse ironized on this spectacle in his first lecture:

> At first glance, Anthropology presents itself as the Science of Man—of man not only dead, but skeletal. Then again, open a textbook of Anthropology. The pages and illustrations are full of dead things. More skulls and skeletons. Then come flints, which, as you advance toward the last page of the volume, take on an increasingly worked-over appearance, but nonetheless stay within the province of dead things. . . . What is this dead and motionless thing? A thing that once supported life. What is this skeleton? The armature of gesture. Those instruments so admirably and scientifically classified? Tools that prolong the human gesture. . . . Gesture, so to speak, is the very mechanism of man. Everything that you can show me in the skeleton is made to act. Man is a complex of gestures. . . . Man is not a skeleton, man is not a skull, man is not a series of tools or of axes starting with chipped stone and finishing with the machine gun.[80]

Though Jousse in this initial lecture made a friendly reference to a book by a colleague in the room, Georges Papillault—who would within a few years be deboned, measured, and reported on to the Anthropological Society by another colleague, George Montandon[81]—his aim was transparently to redefine the object of their study, "man," and to downgrade the corpus of evidence by which man had been made knowable, in the terms of late nineteenth-century knowledge. It was a strange way of taking up residence in a school that would continue to house his lectures until 1951, printing on each annual course list the tag: "The anthropological researches of M. Marcel Jousse have as their aim to seek a liaison among psychological, ethnological and pedagogical disciplines."

Considering its history, the École d'Anthropologie was a strange place to find the Jesuit and promoter of the Palestinian Rhythmic Recitation. Its founders were

> neo-Lamarckian "materialists" who had long believed that human phenomena were the result of material interactions alone, that physical and sociocultural evolution thus correlated, and that evolution was linear and progressive. It was as materialists and freethinking republicans that they had been drawn to anthropology in the first place, seeing in this "new science" an opportunity to prove what they knew to be true about human development, and at the same time to use this science to benefit humanity.[82]

The skulls in the entryway stood as mute witnesses to this scholarly agenda, their cranial capacity minutely recorded by Paul Broca and his successors in the effort to compile as complete as possible an archive of human diversity, from the "lowest" to the "highest" races and types. Papillault, the anthropometrist singled out for recognition in Jousse's inaugural lecture for having mentioned "gesture" in one of his books, specialized in "biopsychology," or the "bio-psycho-social concatenation." "In reality," he affirmed, "social phenomena are merely the extension of biological processes."[83] The sociology of Durkheim and Mauss was oblivious, he thought, to the fact that "the most widespread form of organization, the one most independent of environment and race, produces different results, yields profoundly different outcomes, according to the biological characteristics of the individuals that make it up."[84] As he saw it, no universals applied to the human race except biology.

What was Marcel Jousse doing in this company? Those "immense ethnic laboratories where the Psychology of Recitation . . . discovers with increasing interest the great universal laws of the Oral Style" had, despite their differences of language and culture, a common articulation with the human body. The "human mechanism" operated through reflex, memory trace, and "replay" (*rejeu*). The more the conscious mind could be excluded from the process of memory-inscription, the better: "The mimetic and affective character of one phase of a propositional gesture often triggers, *almost automatically*, another propositional gesture that comes to counterbalance the first as its synonym, antithesis or consequence." "Thus we hear the Reciters triggering one pendulum-swing [i.e., verse] by another, exactly as if by *reflex* action."[85] "Man is an interactionally and bilaterally mimetic animal." "The law of Mimesis cannot pour itself out otherwise than in conformity with the human form."[86] If oral literature delights in symmetries and parallelism, this is because the humans who vehicle it have two hands, two feet, and swing them alternately as they walk. The desire that shows clearly through such passages is to give cultural forms the inescapability of gravity or geometry.

If Papillault's explanations were reductive in the eliminationist sense, seeking to clear the decks of the poorly substantiated (as he saw them) nonnaturalistic theories of culture and society, Jousse's were aspirationally reductive, seeking to encode cultural information in a wholly bodily language. In the terms of Ribot's law, often invoked in Jousse's statements on memory, the ideal of mnemonic training would be to inscribe the message to be preserved on the earliest-acquired, most primitive, least reflective layers of the psyche. The Word became flesh, in the world of Palestinian rhythmo-pedagogy, in that it was the product of stimulated muscular tissue reverberating in answer to prior reverberations. Scientifically speaking (and Jousse's idea of science was that of nineteenth-century positivism: observation, measurement, comparison), to account for the Word through the flesh was possible, as the flesh and its behaviors could be understood, the Word not quite so easily; to account for the flesh by the Word was the futile task of centuries of preaching that Jousse left to one side.

The flesh that preserved and transmitted the Word was through-and-through mimetic, reproducing phenomena in gestures that other bodies then imitated. Moreover, this mimetic body was located in a general model of "energetics" shared by physicists, biologists, social theorists, philosophers of art, and cosmologists in the years around 1900.[87] In a universe heading inexorably toward disorder, why should anything be permanent? More specifically, why should anything in such a universe be remembered or repeated? The body that Jousse took from this tradition functioned as an "accumulator of energy," a reversible battery that gathers energy from its environment and dispenses it in actions upon that environment.[88] A vibrating medium in its own right, the body communicated its energetic frequencies to neighboring bodies. It was thus autographic. Frontiers between science and art dissolved: the symbolist aesthetician Jean d'Udine saw in the vibrations of colloidal tissue the root explanation for heredity, synesthesia, and the effect of artworks on their observer.[89] Thus Aristotle's observation was confirmed at the cellular level: "Man is by nature the most mimetic of all animals" (Poetics 1446a). The body invoked by the semioticians of rhythm was exactly what they needed it to be.

"Four Obscure Jews"

Collective authorship means that no act of composition is an absolute beginning. To speak is to recite, to cite again. The New Testament recites the Old; Mary or Elizabeth recite Hannah; Christianity recites Judaism. A barrier may have been set up between the Old and New Testaments, a barrier consisting precisely of the concept "new," but in oral tradition, with its "bringing forth old things and new," no such difference can be maintained.

When Jousse insisted on the Jewish-traditional character of the *Magnificat*, he was surely awakening in many of his readers a memory of the disdain of the most prominent of the antimoderns, Charles Maurras, for the leveling message of the Gospels. Although professedly agnostic, Maurras was proud to call himself "Catholic" in the sense of belonging to a political-aesthetic tradition of order and hierarchy. As he said (before censoring his own text) in 1894:

> The chain of ideas I set forth here is quite sufficiently pagan and
> Christian to deserve the beautiful title of Catholic which belongs to
> the religion in which we were born. It is not impossible that along
> the way I may have brushed against this or that passage of the Bible,
> but I would hardly be able to say which ones. Intelligent destinies
> have caused the civilized peoples of the south of Europe to be barely
> if at all acquainted with those turbulent Oriental scriptures, except
> as truncated, recast, transposed by the Church into the marvel of
> the Missal and the whole Breviary; that was one of the philosophical
> dignities of the Church, as it was, too, to set the verses of the Magni-
> ficat to a music that softens their venom. . . . Never will I abandon the
> learned procession of Fathers, Councils, Popes, and all the great men
> of the modern élite to cleave to the Gospels of four obscure Jews.[90]

Highly placed clerics shared Maurras's low regard for the Jewish origins of Christianity. Jean-Baptiste Frey, of the Holy Spirit Fathers, secretary of the Pontifical Biblical Commission, took to the columns of the Commission's house journal, the *Revue biblique*, in 1915 to reprove a rumor recently given new circulation in an article by Adolf Harnack. German exegesis, true to its tendencies, had "mauled" and discredited the Lord's Prayer as given in Matthew and Luke. "But all these attempts to strip the Lord's Prayer of the divine seal which Jesus Christ imprinted upon it and to cheapen it to the level of a worthless imitation," insisted Frey, "are without effect and in the end cast only a brighter light on its transcendent beauty."[91] The phrases "hallowed be thy name," "thy kingdom come," "thy will be done," had been declared inauthentic, "with the serious drawback of bearing a too pronounced Jewish odor, and credited to the author of the first Gospel."[92] As a consequence of this "destructive criticism," "the Lord's Prayer is deprived of all merit, because it lacks originality: it is nothing more than a mosaic of formulae that had been in long use in Jewish assemblies. There is nothing Christian about it; it is Jewish in both form and substance." Harnack's "audacious affirmation" had been taken up by Protestant theologians, but "the Jews distinguish themselves, in this attack on the Lord's Prayer, by a singular vehemence."[93] The "ancestor, or at least the close relative," of the Lord's Prayer is specifically "the Kaddish, a sort of doxology recited in synagogues at the end of the Haggadah reading." Despite some inescapable verbal parallels, Frey insists on the *distinguo*. "Even if the letter is the same, the spirit is different. The rabbis who rounded out the Kaddish knew what meaning to give the words 'hallowed be thy holy Name' or 'thy kingdom come,' and when asking for the deliverance of Israel, the restoration of the Jewish capital, and the prosperity of the Jewish nation, they were consciously keeping to the original tone."[94] The scope and content of the Lord's Prayer, by contrast, is entirely other: the reign of justice, mutual forgiveness, universal brotherhood. "Thoughts of so great a moral elevation can be found neither in the Kaddish nor in the other prayer formulae of Judaism. . . . The simple words, 'Forgive us our trespasses, as we forgive those who have trespassed against us,' cannot receive a *Jewish* interpretation; the idea is specific to the new religion, and this fact specifies well enough the difference of level between the two prayers."[95]

For Frey, then, to trace a Christian text back to its Jewish origins is a gesture with blasphemous implications. Whoever would do so, Protestant, modernist, or Jew, would be guilty of the archetypal Jewish error of confusing the "letter" with the "spirit, which alone giveth life." The oral style recognized no such barrier, and indeed its exponent went far in the other direction, situating Christianity within Judaism rather than insisting on their incompatibility. Jousse answered Frey's question, "Is the Lord's Prayer Jewish or Christian?" thus: "The Lord's Prayer can be recited by a Jew as a Jewish prayer because it is made entirely of Jewish formulae; it is Jewish in its grouping of formulae, and nothing in the new juxtaposition of formulae that Rabbi Yeshua made can disturb a Jew."[96] As a reciter in the tradition of Torah, Rabbi Yeshua could hardly have done otherwise.

But as a cleric in the interwar Catholic Church, Jousse certainly could have done otherwise, and many of his coreligionists did. Maurras's movement, the Action Française, dominated conservative (and therefore, Catholic) opinion in the first half of the 1920s. Founded in the wake of the Dreyfus affair, the Action Française, nationalist and monarchist, took as its mission to save the country from "the selfish petite bourgeoisie," "Protestant liberal-democrats," "Socialist collectivism," "the money power," Freemasons, and "the new religion of human rights," all of which, it charged, had left the Third Republic in a state of "irresponsibility."[97] "Down with the Jews!" was of course one of its rallying cries.[98] The Action Française was also emphatically on the side of the Vatican in the Modernism crisis and the 1905 separation of Church and State. At the same time, varieties of Social Catholicism that had developed in the circles of Marc Sangnier (Le Sillon) and Maurice Blondel (La Semaine Sociale) came under ecclesiastical censure.[99] (In December 1926 Action Française, too, was condemned by the Vatican, a move that took most French religious conservatives utterly by surprise.[100]) A group of clergy and lay people, calling themselves the Friends of Israel, campaigned against such expressions of traditional anti-Semitism as the prayer "for the perfidious Jews" in the Good Friday liturgy, and were disbanded by order of the Holy Office.[101] The group had existed for thirteen months. The Catholic convert to Judaism Aimé Pallière provoked much shaking of heads with his judgment that "Jesus was simply the last of the Prophets"; the Jesuit journal Études traced his apostasy to Modernism.[102] The investment in keeping Judaism and Christianity separate was so visible among the very people who were most receptive to Jousse's claim to have dissolved the Modernist problem that it is almost paradoxical to see them countenancing the oral style's practice of fusion between the Testaments, of which the consistent use of the title "Rabbi Yeshua of Nazareth" is the most visible sign. So when the organizing committee of the 1927 Semaine des Écrivains Catholiques (Weeklong Festival of Catholic Writers), dedicated to the general theme of "Our Catholic intellectual renaissance and the restoration of the Christian world," held a panel on the Jewish question, it provoked strong reactions.[103] Paul Halflants reported:

> Anti-Semitism had a rough time of it at the session of the Week of Catholic Writers in Paris, on the evening of December 9. . . . A Jesuit turned to M. Edmond Fleg, a notorious Jew invited to the meeting to present the point of view of his coreligionists, and unexpectedly uttered on his own behalf the words of the crucified Christ: "Forgive the Catholics our old resentments; we did not know what we were doing." M. Stanislas Fumet, substituting for M. Jacques Maritain as session chair, provoked justified cries of protest by declaring that the

Jewish and Christian religions are essentially one, since they worship the same God. . . .

M. Edmond Fleg is the winning representative of a new literature, specifically Jewish and in no way hostile to Christianity. . . . He is known, moreover, to dream of the religious reconciliation of Jews and Catholics. His attitude at this meeting, where he must not have expected to be so well received, makes this writer dream of his complete adherence to the true Messiah foretold by the prophets.[104]

The high point of the evening was the presentation by Father Marcel Jousse, the famous discoverer of a new psychology of language. . . . He put such energy and conviction into his thesis that we must explain the Greek text of the New Testament through the corresponding Hebrew terms that part of the audience, taken aback, asked if this meant that the Catholics had always misunderstood the words of Christ. Father Jousse reassured the fearful: from the first centuries on, the Church has always seen the danger that the meaning of the divine words could be deformed by their transposition into Greek or Latin molds. And the Church's great labor against Gnostic and other heresies has always been to return the texts to their authentic Jewish meaning.[105]

But Jousse's own texts always insist on the deformation of the Hebrew-Aramaic original by its Greco-Latin gloss. It is just within possibility that a commentator from the reactionary edge of the Catholic cultural landscape of the time has the solution:

> We have already mentioned the scandalous session of the Literary Congress, the so-called 'Week of Catholic Writers,' a singular zoo of scribblers of every stripe and persuasion, mainly interested in being on stage. . . . Naturally, this Noah's Ark was the setting for the stupidest possible pronouncements. To the point that you might question our testimony; but we have our witnesses.
>
> Canon Paul Halflants . . . had to intervene personally at the end of the session, according to M. Maurice Vaussard, to reduce to more acceptable proportions Father Jousse's latest verbal enormities.[106]

For reactionary Catholics, then, Jousse was an agent of the Jewish International. But what did Jews at the time have to say about his reconstruction of the common Jewish-Christian past? Was his oral Talmud at all recognizable to them?

To some, at least. A handful of articles by Étiennette Boucly, expressing unrestrained enthusiasm for the "discoverer of the oral style," appeared in Jewish periodicals in the early 1930s.[107] Consisting mostly of a
diligent summary of the main points of Jousse's teaching, they at times
address their specific audience. "Isn't it in the interest of all civilizations,
of all religions, practically the duty of everyone, to proclaim Israel a
classic author and start to study it as such, in its living, harmonious, and
intellectual life?"[108] "Israel, classic author"—the collective author given
a singular name, the better to stand alongside Homer, Racine, Goethe,
and Shakespeare. Like Martin Buber, also venerated in the pages of the
same periodical, Marcel Jousse will show assimilated European Jews
how to be traditional Jews again, only through recitation and bodily
movement.

> "A Hebrew gesture to go with Hebrew speech! We have lost Hebrew
> mimicry," complains Bialik.
>   Recovering Hebrew mimicry is simply a matter of becoming
> limber enough to let the modeling energy of these Mimèmes radi-
> ate through one's body and hands—those Mimèmes that once were
> wound up in contact with the Palestinian Real and are still ready,
> face-to-face with that same Real, to replay, in unison, the same
> Hebraic articulations of sound.
>   Prodigiously concrete, this Hebrew mimicry, to be sure, but also
> deeply intellectual and pedagogical, because it is the bearer of the
> most awe-inspiring and indispensable Truth. . . .
>   How is it that it took a recent anthropologist to uncover, expose
> and scientifically analyze so characteristic an expressive mechanism?
> What reasons caused Israel, little by little and finally completely, to
> forget?
>   Christians have an easy answer to this question.[109]

Jousse's return to ancient recitation-forms is for Boucly the occasion
to remind educated, anxious European Jews of 1934 (the audience of the
journal *Cahiers juifs*) that they too have a return to accomplish. Bialik
is the poet who contributed most to the revival of Hebrew as a modern literary language, and Boucly makes Jousse answer Bialik's request:
from Rhythmo-Mnemonics the new, returned Jews can learn to supply
the missing gestures.

Twenty years after Jousse's death, the poet, translator, and critic Henri
Meschonnic charged him with "anti-Judaism" and dismissed his work as
"an edification of Christianity." "His capital discovery ends by being reductive." "The anti-Judaism is the accompaniment of a Church. Jousse

is constantly bringing back 'the teachings of the Targoums' to Christian doctrine. . . . Speech is a *trinity*: 'The Speaker, the Speech and the Breath' . . . on the familiar pattern, which is that of Christian theology and of the sign."[110] So long as Jousse remained under Catholic auspices, it seems, his philosemitism could only be decorative or hypocritical. A hard rule; and it seems to disregard the insistent desire to explain away the more metaphysical parts of Christian doctrine with the aid of a certain view of the Targums, rather than to turn the Targums into Christian prophecies. Jousse repeatedly denied that he wanted to convert Jews.[111] But that is not the point. The point is, rather, that what the oral style produces in Jousse is certainly at the end no longer simply "Christianity" as people of 1925 or 1950 knew it, but a mixed, empirical creed. Such a hybrid is sure to displease the Meschonnics as well as the Colmets, and to speak most clearly to those who, like Boucly, appropriate it with something else in mind.

Gallo-Galilean Civilization

After the fall of France in June 1940, the oral style constituted a one-man resistance front.

Just as it had done with Judaism and Christianity, or with rhapsodes and poets, the oral style, in Jousse's thinking, drew new and confusing boundaries across well-known and seemingly inescapable ones. The real war was not between fascism and democracy. Rather, on one side were arrayed the powers of writing: the Roman Empire, Mussolini, Hitler, the Front Populaire government of 1936, Maurras and his ideology of "order." On the other, the ancient Gauls, Vercingétorix, the Palestinian Reciters, Rabbi Yeshua, the peasantry.

> Three years ago, the gifted Duce of Fascist Italy threw out to the four winds the historic pronouncement: "France and Italy are not sister nations." Now there is a sentence, dropped from specially authorized lips, that I would like to see branded on the forehead of every young Frenchman who sits on the benches of junior high school. A French organization had taken as its slogan the inverse of this sentence: "We are Latins. We must follow the Latin path in guiding all our forces."
>
> It is time to recognize that we have been fooled by these men and we are about to see why. The cause dates back two thousand years. . . .
>
> The Gauls are not a Latin people, but when they saw themselves defeated [by Julius Caesar] they threw themselves down on all

fours before the victor. They spoke his language. They imitated his gestures. . . . And this is what they teach us in junior high school under the name of "Greco-Latin formation!" This is one of the most monstrous things that can be imagined in a country. . . . Well, we say No. And when Maurras tries to lead us in the name of his Latin sister toward a destiny that is not ours, we say, "We peasants are not going to follow."

The GAULS had a formidable CIVILIZATION and this civilization was not written, it was oral, and the present speaker was the first to reveal to the world the value of these formidable civilizations that do not know and do not want to know how to write. . . .

After [defeat at] Alesia, the Gaulish aristocracy immediately declared itself Roman. They rushed to acquire, not only the language, but what is more, if I dare say so: stupefaction through Rhetoric. . . . Voltaire! Robespierre! . . . And that went on from 1789 to 1940 with the same mechanisms. . . . You see what we had as our ideal in France in January 1940! And it ended up as it was bound to end up. . . .

But while the Gaulish aristocrats were groveling before the Romans and their rhetoric, the Pagani, [the peasants], those who lived in the "pagi" [the fields], remained faithful to their old traditions as best they could.[112]

This rejection side by side of the Action Française and the Front Populaire expresses resentment no doubt widespread in the occupied zone of France in 1941. But it does not correspond to any of the political options at the time. Pétain had put the surviving members of the Front Populaire government on trial at Riom for having caused the defeat of France; but Pétain's Vichy government was packed with members and sympathizers of the Action Française. If anything, Jousse's championing of the peasantry echoes the Vichy motif of a return to the "real France," but this too was a Maurassian theme, the distinction between the "legal nation" and the "real nation."[113] Jousse's antipathies are all over the map and partly self-contradictory, in anyone else's terms.

The Druids, those great teachers, are comparable to the Palestinian rabbis! Here are our ancestors. . . . But neither the Druids nor the Rabbis wrote. . . . What are the Druids? Teachers who did not write, not preachers, nor orators, but Rhythmo-catechists who transmit memorizations. . . .

The young people who come to my lectures always ask me to analyze Rhythm. Rhythm? We learned it from our mothers' lips and that's where it comes from. Gaulish rhythm came also from another source: from those great Galilean peasants who, at a certain moment, came to interfere with the Gaulish mechanism, so that the Druids of Gaul and the Palestinian Rabbis converged on the lips of peasant mothers.

For two thousand years now, peasant mothers have rhythmo-catechized their children, that is, repeated to them in echo what they learned from their mothers and grandmothers. . . . We are not Greco-Latins. We are conscious and learned Gallo-Galileans, and when we appear, the auditoriums of our country are instantly filled.[114]

The débâcle belongs not to the peasants of France, but to bookish civilization.[115]

Those self-negating Gallo-Romans led naturally to 1789 . . . The petits-bourgeois took up the thread in 1789, they maintained the Gallo-Roman mechanism and this naturally came to a conclusion in 1940. So don't be surprised if the anthropologist who is speaking to you today has only one mission: to abandon the deceptive Gallo-Roman mechanism and to keep *pure* the Gallo-Galilean mecha-nism. . . . We are pursuing a total renewal and that is why the young understand me best, because they feel, not the call of the race but the call of the Tradition, the call of the Gallo-Galilean.[116]

Many Frenchmen took advantage of the defeat of 1940 to assign blame and to call for "national renewal" conditioned on one or another kind of "purification."[117] Jousse renarrates the history of the past two thousand years to make the present war an episode in a longer war and to forge an emergency alliance among Druids, Rabbis, and Mothers, the resistance network of the oral style. Defeat inspires an imaginary overthrow of writing by the underground mnemonic front. Exorbitant, desperate, its draw-ing of battle lines presages the polarized world of McLuhan's early media theory with its efforts to "explain" the behavior of Africans, Asians, Soviet commissars, and other Others on the basis of their allegedly shared oral mentality.[118]

And it is to the Oral Front, not to the confessional, that Jousse calls his Jewish hearers. The plurimillennial reign of writing will come to an end; orality and gesture will emerge from the low places and be raised high; the mothers, peasants, Druids, and Rabbis will triumph over the Legions, the scribes and the rhetoricians; it will be an apocalyptic reversal of the media world, the *Magnificat* with its venom intact.

5 /   Embodiment and Inscription

*The knowledge that the first material on which the mimetic faculty tested itself
was the human body should be used more fruitfully than hitherto to throw light
on the primal history [Urgeschichte] of the arts. . . . Perhaps Stone Age man
produced such incomparable drawings of the elk only because the hand guiding
the implement still remembered the bow with which it had felled the beast.*
                                                            —WALTER BENJAMIN

Oral tradition is not the antithesis of writing, but a particular kind of
writing, an inscription on other human minds. Ever since Plato's Socrates
dismissed writing on physical surfaces as a "trick" or "game" (ἐν μὲν τῷ
γεγραμμένῳ λόγῳ . . . παιδιάν . . . πολλὴν), the better to praise the "true
writing on the soul concerning justice and beauty and goodness" ([τὰ]
τῷ ὄντι γραφόμεν[α] ἐν ψυχῇ περὶ δικαίων τε καὶ καλῶν καὶ ἀγαθῶν),
this second writing has always been understood as a metaphor—the
writing that is precisely not writing, but consciousness.[1] It could only be
a metaphor, goes the conventional understanding, because the soul is a
nonphysical thing; no one could possibly make a mark on it. But just as
the broader term "notation" enabled us to see around certain supersti-
tions of exactitude in alphabetic writing, so the term "inscription" may
encompass both kinds of writing so named by Plato, the unironic writing on paper and the ironic writing on the soul. It is worth insisting that
"inscription" as used here is not merely an equivocation or a metaphor,
but one more effect of the perturbation brought by oral tradition to the
categories of literature. For the customary stark division between orality
and literacy must collapse, taking with it some of the traditional privi-
leges of the human, once the transmission mechanisms of oral tradition
are moved to the other side of that erstwhile line. It is not that all differences vanish, only that the relative differences that come to replace
the stark antithesis of writing and nonwriting put us in a better place to
describe the relations among the materials and occasions of these different types of writing.

Materials Science

Oral tradition has in common with all forms of inscription to be conditioned by the materials with which it works. The method of inscription is not the same for wax, parchment, magnetic tape, photographic film, papyrus, stone, or human memory, but some analogous operations must occur in each case: a surface to be inscribed (the material support) is prepared and an inscribing device is deployed in such a way as to leave marks on the support, marks that later can be read or (what is the same thing) copied or deflected onto a future support. Operations that result in a successful inscription with one kind of material support might have no effect on another: consider the effects of applying to silver nitrate film the inscription-method invented to take advantage of the properties of marble, or the converse. Likewise, erasure, the negation of information, will take a different form for each material support. Fire annihilates paper documents but preserves libraries written on clay. What counts as writing is a philosophical matter; how to do it, a practical question that starts from the material support.

A reiterated tradition, we might say, must leave marks on the souls or minds of its hearers, insofar as their successive readings or recitals reproduce these marks. What are the properties of human minds that lend themselves to iterable inscription? In other words, what is memory? Until the workings of the human brain have been adequately accounted for, the explanation of memory will have to start from observations of gross empirical practice, thus the question may be reframed: How have humans performed acts of inscription on the minds of their fellows?

A child recites, "Ring around the rosy, / pockets full of posy": we know that children, even preliterate children, have been doing something like this for generations, so some kind of inscription and transmission must have taken place, involving mouth and ear and text and eye.[2] Reasons for the routine's transmissibility include the number of "cues," as David Rubin calls them, that bind the elements of the ditty together and increase the likelihood that they will be repeated in facsimile rather than dissolved into the generality of paraphrase: there is rhythmic pattern,

RING aROUND the ROSy,   / - / - / -
POCKets FULL of POSy   / - / - / -

there is alliteration,

r . . . r . . . r . . . zy
p . . . p . . . zy

there is rhyme,

rosy
posy

there is a tune, and there may be a circle dance with up-down movements of hands and the impact of feet on the ground to reinforce the patterning of the words.[3] In all this there is selection and combination from the indeterminate possibilities of language in general, magnetizing, as it were, the sounds and meanings of the composition's words into a highly organized whole. Not all of the organizing is supplied by this specific text. A child who has grown up in a milieu of English speakers has been prepared by thousands of experiences to recognize /p/ as a phonemic entity with systematic relations to other English phonemes. He or she will have come to treat as perceptually salient or purposeful the alternation of stressed and unstressed syllables (speakers of Japanese or ancient Greek would not consider this difference as metrically important) and to recognize a rhyme as a design feature rather than an irritating recurrence (ancient rhetoricians taught their charges to use sparingly the jingle of similar word endings).[4] The learner's mind, in other words, has been prepared—habituated—to receive and give attention to the organizing devices usual in English-language rhymes and songs. This habituation is an extension of the general competence shared by speakers of English. (Some of these devices are also found in other languages, but not necessarily in the same combinations or with the same prominence.)

Put in terms of communication technologies, the acquisition of familiarity with poetic devices is a formatting of the receiving surface, a preparation of that surface to receive information of a particular consistency. The transmitting surface of the poem or song is patterned with a corresponding set of features so as to favor the receiver's retention of the details of the text—details as many and as fine as possible, where accurate reproduction is valued. When the inscriptional support registers these features, we have textual memory; when the receiving surface becomes in its turn a transmitting surface, and the listener repeats the song (perhaps with variants) to new hearers, we have an episode of textual tradition. The variants, if variants there are, may exhibit similar features of memorability. One may assume that other things being equal, the more densely the text is packed with pattern, the more securely it takes up residence in the prepared mind.

Inscription on human minds, as it has been going on for thousands of years, implies an at least implicit knowledge of the material with which it deals: how to make an inscription, how to transfer it to a new surface, how to verify it, what tends to go wrong and how to compensate for error. The techniques will vary with the kind of text to be inscribed and the value of its content to the community in which it survives. A counting-out rhyme will have one kind of structure, an epic recital lasting several days will have another, a joke yet another. The "posy" rhyme exhibits several mechanisms for mental inscription. Another example will show how and why these mechanisms vary in application. I possess word-for-word the first lines of the counting-rhyme

Thirty days hath September,
April, June, and November,

though an absence of definite cues may make me pause at "June" (is it "June" or "May"? Both have one syllable). The calendar will be intact if I can assign those four months their thirty days and no more. But going on, the cues weaken, and different people will remember differently the remainder of the ditty, which winds up more or less thus:

All the rest have thirty-one.
Except for February, which has twenty-eight,
and twenty-nine in a leap year.

"All the rest . . ." is bound to the foregoing by similarity of meter, but with "February" meter gives way to unbound prose and the anecdotal features of an exception. Proportionally, February is a more narrowly localized problem for astronomers and check-writers than the other four months that lack a thirty-first day, so perhaps mnemonics here obeys a rule of efficiency. Form is meaningful here even where it compromises meter or meaning. If the rhyme is meant to regulate the calendar, the hesitation over "May" and "June" shows a conflict between rhythmic indifference and semantic distinction, and "February," standing outside the regularities of both calendar and rhythm, demonstrates its exception by making the pattern break down.

To say that the inscriptional support correlates with features of the text to be transmitted is not to say that transmission is problem-free. But the types of error in transmission are keyed to different types of convention: although it must be rare for anyone to hear an N as a U, the similarity of handwritten *n*'s and *u*'s is familiar to any scholar who has worked with manuscripts. Chinese scribal errors are different in kind and scope from, say, Latin ones, and both of these differ in structure from the kind of binary error that can corrupt a computer file.

If oral tradition is an inscription on human minds, its conditions of success will have partly to do with the formatting imposed by previous culture, partly with the readiness of individual subjects, partly with the properties of the texts to be transmitted, and partly with the capacities of the human brain—all these characteristics in interaction and mutual accommodation. It has been common knowledge since the 1880s that organized and meaningful material is more readily memorized than random sequences: Ebbinghaus advanced a proportion of ten to one for the difference in retainability of meaningful and meaningless text.[5] Meaningful text with a maximal organization of linguistic and thematic features keyed to the habits of the audience (patterning of phonemes, syllabic regularities, melody, rhymes, stanzas, formulae, topoi, analogies, antitheses, coherence of action, and the like) should then have a much higher rate of retention, especially in the case of hearers who make it their profession to learn and repeat such texts: the bards. A text designed for oral transmission will be the product of all these requirements, adapted in an ongoing fashion through its recitals to the kinds of formatting that its hearers have already internalized. To quote again the observation of Walter Benjamin: "To an ever greater degree the work of art reproduced becomes the work of art designed for reproducibility"—and this already under preindustrial conditions.[6] Homer's workshop, we may say, devised

> what many would feel to be the *Iliad*'s most significant structural device: the use of narrative time to structural effect in a purely *formal* way, by the establishment of artificial patterns of symmetry and closure in the arrangement of events. A famous instance is the elaborate mirror symmetry between the first and last books: the ten- and twelve-day suspensions in each are part of a wider pattern of correspondence taking in the two initiatives from Apollo, the complementary scenes between Thetis, Achilles, and Zeus, and the symmetrical restitutions of children to suppliant fathers. Patternings of this kind continue to dominate or monopolize discussions of Homeric "structure"; but it is important to stress again that, while such structures clearly do encourage the reader to anticipate the closure of large-scale patterns in the narrative, they are no part of the poetics of classical plotting. Indeed, it is moot whether these non-causal devices for the structural shaping of readers' narrative expectations really qualify as "plot" at all in any vernacularly understood sense.[7]

By the time "the poetics of classical plotting" were articulated (most notably by Aristotle), these formatting devices no longer served a functional purpose, as inscriptional surfaces other than human memory had been found useful and reliable for recording epic, tragedy, and narrative. "All the temporal devices surveyed here help the reader to construct, from what presents itself as a purely linear narration of events, an increasingly confident global model of the shape and structure of the story as a whole."[8] If we were epic bards, we would, as we heard it, be "constructing" the poem in a different sense of the word, not merely "construing" it but building it, and making not merely a "model" for interpretation but the groundwork of the version of the poem that we, on the next occasion, might sing. Rhythm is schematic and projective.[9] It sets a pattern and supplies the elements for filling it out. The potential text and the realized text are not one, and the writing on human memory is precisely a marking of potential features (goals, means, areas of indeterminacy). A text that is made to be learned trains its hearers in the mental competencies needed to reproduce it, in a looping process of coevolution.[10]

Thus, the "laws of epic narrative" proposed by Axel Olrik in 1908 can just as well be turned around and considered as discoveries about the properties of the human minds on which folkloric narratives are inscribed.

> *The "epic laws" of oral narrative composition.* In popular narrative, storytellers have a tendency to observe certain practices in composition and style that are generally common to large areas and different categories of narratives, including most of the European narrative tradition. The regularity with which these practices appear makes it possible for us to regard them as "epic laws" of oral narrative composition. . . .
>
> The narrative does not reveal a multiplicity of features corresponding to real life. It has fewer characters and, above all, fewer forces are simultaneously brought to bear on the fate of a human being. If something must be depicted as a result of many concurrent forces, the narrative brings them into play one by one.
>
> Among others, the so-called Law of Two to a Scene is a manifestation of the clarity of the narrative: the narrative only reluctantly brings more than two characters on stage at the same time; under particular circumstances a third (subordinate) character may be added for a short performance. . . .
>
> The narrative schematizes its characters and incidents, i.e., gives them only the features most necessary to the plot. Several similar characters or similar episodes are intentionally made to resemble each other as much as possible. . . .
>
> Oral narrative composition does not know detailed description . . .
>
> Normally, repetition is accompanied by progressive ascent: the hero fights three times, with each successive fight more difficult than

the previous one; or he makes three attempts, and he succeeds only the last time . . .

Any ability of a character or thing must be expressed in action; otherwise it has no importance for the narrative. . . .

The narrative preferably places simple contrasts together: big against small, man against woman, human being against animal. The contrast itself must characterize the special nature of each. . . .

When two characters appear at the same time, the narrative will establish a contrast in character between them, often also a contrast in action. . . . *The Law of Twins.* When two characters appear in the same role, they are both depicted as being weaker than a single character.

The Law of Three. The narrative has a preference for the number three in characters, in objects, and in successive episodes. . . .

The Law of Final Stress. When several narrative elements are set alongside each other, emphasis is placed on the last in the sequence: the youngest of three brothers, the last of three attempts, etc. The epically important character is normally in the position of "final stress."[11]

Olrik's "laws" somewhat empirically collect notable features of folktales that correspond to articulations of the human semantic space, elements of the formatting common to minds and texts under the conditions of oral recording and transmission.[12] With allowance for differences in genre, they can be assimilated to features of the "oral-gestural style" documented by Jousse, in his polemic for the fidelity of the spoken word against the fragility of script. Each of Olrik's "laws" can also be described as a trigger activating one or more "cues" to mark out the course of the story: If one brother is shown failing to defeat a giant, two more brothers must be in the storyteller's sack. If a story begins with a man and a bear, it will continue in a way that draws out the differences between human and animal. And the law of "schematization" excludes from the story more details than the teller can presently use; but an exceptionally able teller may be able to use more than an average teller can, and thus put flesh on the schematic bones that suffice only to get the tale retold.

The formal features of orally conveyed texts are often described as limitations. The folktale as characterized by Olrik is certainly poor in psychology; Parry portrays the epic poet as limited to a narrow scope of time, as a result of which he must use stock phraseology; Goody and Watt argue that historical consciousness cannot develop in a textual economy where the past subsists only insofar as it serves a present need. But if oral traditions are constrained, it is chiefly by the capacities of the human mind, and they "are stable, in large part, because of the combinations of these constraints. . . . Combined constraints produce effects much larger than those of the individual factors by decreasing the memory load and increasing the number of cues to recall."[13] What is best remembered will be what is overdetermined, cued by multiple networks of causes. The art of the writer who writes on human minds consists in drawing the greatest complexity from the limitations of the art form's material support.

Techniques of the Body

Oral epic—and other genres of oral composition—is thus a technique of the body, in the sense defined by Marcel Mauss.

Mauss's famous article, originally delivered as a talk in 1934, begins with a series of anecdotes from experience. The way Mauss was taught to swim, around 1880, was no longer current in 1930. The human body had not changed, the specific gravity of water was still the same, but a new technique had been discovered or imported, and that was what the children of 1930 learned as "swimming." Another story: A British regiment received, in recognition of their heroism during a certain battle, a French brass band. But when the regiment tried to parade with its new band, "the results were discouraging . . . the regiment had kept its English way of marching and now it was following the French rhythm. . . . The luckless regiment of tall Englishmen was unable to march."[14] A difference in the execution of the most basic of human techniques, walking, hobbled this venture in international understanding. Another time, Mauss was hospitalized in New York. "I wondered where I had seen young women walking the way my nurses did. I had time to think about it. Finally, I realized it was at the cinema. When I came back to France, I noticed, especially in Paris, that this style of walking was becoming widespread. . . . American ways of walking were coming to us through the cinema."[15] Mauss saw in such anecdotes the demonstration that "these 'habits' vary, not only according to individuals and those who copy them, but especially according to societies, types of education, conventions and fashions, and levels of prestige. We must see in them the techniques and the product of collective and individual practical reason, whereas the usual explanation calls on the mind and its faculties of repetition."[16] These techniques need to be "installed" (*montées*) in the body, they do not derive immediately from its physical properties, and it is "social authority" that installs them.[17] Mauss reminds us that "the body is mankind's first and most natural tool. Or rather, if we would rather not speak of a tool, the first and most natural technical object, and at the same time technical means, of mankind is the human body. . . . Before there were techniques with tools, there were techniques of the body."[18]

The body here is not "bare, unaccommodated"; we see from Mauss's stories about walking and swimming that a "technique with tools" such as the cinema or the brass band can express and transmit, or conflict with, a technique of the body. But whether technique is with or without tools, with or without external materials, technique implies social governance. Human beings, even or especially "primitive" human beings, are hardly imaginable without such an articulation among techniques of the body, techniques of external matter, and social structure. The central role of these toolless techniques is often forgotten. André Leroi-Gourhan advanced in his account of anthropogenesis that "Technique is both gesture and tool, organized in chain form by a veritable syntax that gives to operational sequences both fixity and flexibility. Operational syntax is proposed by memory and is born between the brain and the material milieu."[19] Bernard Stiegler adds:

> Epiphylogenesis, a recapitulative, dynamic and morphogenetic accumulation of individual experience, designates the appearance of a new relationship between the organism and its milieu, which is also a new state of matter: if the individual is organic and thus organized matter, his/her relationship to the milieu (to matter in general, organic as well as inorganic), insofar as it is a matter of a *who?*, is mediated by this organized although inorganic matter that is the organon, the tool with its instructing role (its role as instrument), the *what?* It is in this sense that the *what* invents the *who* just as much as it is invented by it.[20]

Techniques of the body, that kit of "the most *natural technical objects*" that are part of both the *who* and the *what*, tend to be ignored in histories of technology because they do not so easily permit that discipline's characteristic reversal of perspectives, in which the tool makes man rather than man making the tool. "*It is because it is affected with anticipation, because it is nothing but anticipation, that a gesture is a gesture*; and there is gesture only once the tool and artificial prosthetic memory exist outside the body, as if to constitute its *world*."[21] Would not admitting "techniques of the body" to the status of full technical objects perform even more effectively the discordance of stages in the speculative genealogy of the human that Stiegler calls for, as an unquiet reader of Leroi-Gourhan? It would at any rate put the posthumanism of technology studies on a new basis, by neutralizing the differences between self and tool, inner and outer, the technical and the symbolic.[22] Indeed, the study of media founds itself on the story that first there was oral tradition, and then came the "extensions of man," starting with writing; but what if the first medium were a capacity that "man" identified and trained in himself?[23]

"Rhythms . . . integrate the subject in time and space."[24] This may be so for the baby being rocked to sleep, for the log-splitter, for the cyclist hitting her stride at the start of a long climb: a rhythm automatizes action, distributes effort uniformly across many impulses, designates an endpoint and pulls toward it.[25] Rhythm can also be imagined as the background space-time in which the reciter-composer of a long oral text performs the work. A metrical pattern assures continuity. In every line are one or two well-worn phrases that seem to take care of their own articulation. There are, in the farther distance, sunrises, mealtimes, embarkations and disembarkations, segments without surprise or effort beyond pronouncing them. There are passages lit by the memory of another's recital to be emulated and thematic echoes laid out for symmetry and memorability: here perhaps you know what it should be like without knowing exactly how to say it. And in the spaces to be traversed between such known landmarks are areas known so far mainly as strategy: tasks that arise in the course of the story's development (now that I have Diomedes on the battlefield, how will I make his performance memorable?), augmentations, the determination to outdo a rival. For a trained bard, the recurrent and predictable features of the genre take up almost no room in consciousness, so that attention can be focused on the challenges.

In trying to sketch out this map of epic as affective space-time, we must find the image of the book, where every word is spelled out, the very avatar of "misplaced concreteness." The epic in its virtual or potential state, when it is not being recited, subsists as a mental outline or map, we may say; a version of the map is what is "transferred" (a more accurate term would be "emulated") from one competent reciter to another. "The basis of their renderings of the same story is a mental text which they keep and develop in their memory. . . . Every singer must develop his/her own mental text of a particular story on the basis of hearing, learning, and what may be called mental editing."[26] To say that a text like the Homeric epics is preserved in memory is not wrong, but what do we mean by memory in such a case? Alfred Binet, when investigating the apparently supernormal abilities of chess players who could engage in and win up to forty simultaneous games, found that their memory and attention did not engage each separate game as a history of moves, or as a game board loaded with all its pieces. "What makes a strong player? First of all, a great power of combination." "What makes it possible to engrave in the memory a series of moves or a position is the ability to give these moves or this position a precise meaning." In comparison, a novice or average player trying to enter into the thinking of an expert is "like an illiterate person memorizing a line of print." Expert players do not see the pieces as things, nor do they see all the pieces in their mind's eye: they see a complex of forces and seek a way through them. Because a game is "a struggle between ideas," the more ideas one has acquired, the better one can confront new situations: thus Binet frames "a paradox of memory: we lighten the weight of its load by adding to it."[27]

If, in the absence of other material supports, transmission of orally mediated texts is a technique of the body, what kind of a body is it? Rather than impose a model of the body, we might elicit it from the behavior of recited and transmitted texts, on the assumption that the points where they falter or explode in variants will reveal properties of the body-and-brain at work. Texts are not equally memorable from start to finish, and a comparison of versions will show which parts are more volatile. A sequence of events often needs to be held in place by an external cue: the ripening of a certain fruit as signal for the start of a season, or a series of ritual acts for which the community of participant-observers serves as timekeeper and referee. An initiation text collected by Jack Goody has its heroes interrupting a half-executed ritual to ask themselves:

> "How is it that you first carried on
> without finding out
> and now you turn back again
> to get to this point?"[28]

It turns out that the informant, sitting alone with an anthropologist in a tent outside the initiatory season with no informed audience to guide him, had gotten confused about the order of ritual events, but rather than break the frame of the story and "take it from the top," he made his characters bear the burden of canceling the improper ceremonies and starting again from the last good sequence-marker. The reciter is typically situated; in this case it was a relative freedom from the usual situation, a gap in the "script" of repeated activity, that loosened the narrative's pegs. Boundaries imposed by the calendar, the seasons, the social environment constrain information, defend it from randomness. On another level, we may see the ongoing recital cueing itself with aural echoes supplementing the regular mechanism of formula:

> [David] Bouvier, studying the distribution of repeated verses in the *Iliad*, noticed that the interval separating two successive occurrences of a repeated verse is most often fairly brief. That is, a verse which has

not been repeated after a short interval of time has a strong likelihood of never being repeated. The interpretation he suggests, with many precautions to be sure, is the following: while he is executing his song, the bard keeps for a time in memory the verses that he has just uttered, which is apt to make him utter them again; then he forgets them, and the probability that he will repeat them then rapidly approximates zero. The bard, in a word, mobilizes only his short-term memory, that which retains a small amount of information during a brief time.[29]

The repetition of whole verses is not imposed by the formulaic system (the main job of which is to make ready part-verses apt to be joined with other part-verses), so the distribution noticed by Bouvier must be a gratuitous effect of short-term recall and association on the reciter's part, an involuntary feedback superimposed on the vaster memory bank that we call, since Parry's theses, "the system." In these nonfunctional whole-verse repetitions, we can see the feeble mortal struggling with the transpersonal Homeric machine.

The body that recites has limited powers—

πληθὺν δ᾽ οὐκ ἂν ἐγὼ μυθήσομαι οὐδ᾽ ὀνομήνω,
οὐδ᾽ εἴ μοι δέκα μὲν γλῶσσαι, δέκα δὲ στόματ᾽ εἶεν,
φωνὴ δ᾽ ἄρρηκτος, χάλκεον δέ μοι ἦτορ ἐνείη,
εἰ μὴ Ὀλυμπιάδες Μοῦσαι . . . μνησαίαθ᾽ . . .
The number of [warriors] I could never tell or count by name,
not even if I had ten tongues and ten mouths,
an unbreakable voice and a diaphragm of bronze,
if you, Olympian Muses, did not . . . remind me . . .[30]

—and its brain will sooner or later attain an upper limit, say in the nesting of narrative levels or its ability to keep track of a large cast of characters: Olrik's laws are an acknowledgment of the brain's retreat to a zone of semantic predictability and simplicity, of lesser effort. But the "epic technique-of-the-body of oral verse-making" (to force together Parry [1930] and Mauss [1934]) has evolved means of compensating for the scanty resources of amateur or folk recital. In a tradition maintained by somatic transmission of memory cues, ongoing "bricolage" or "dream-work" is the order of the day, and even forgetting can generate a compensating network of new clues with a different pattern of coherence, as Bartlett and Freud found in their experiments with memory.[31] The more we look for clues and structuring details, the more we will be inclined to credit Homer with the ten mouths and tongues, the unbreakable voice and the guts of bronze—the tools and techniques—that he swore were not part of his media universe.

Merleau-Ponty distinguishes the "objective body" (as seen from the outside) and the "phenomenal body" (as experienced from the inside).

> The subject placed before his scissors, his needle, and his familiar tasks does not need to look for his hands or his fingers, because these are not objects to be found in objective space, bones, muscles, nerves, but powers already mobilized by the perception of scissors or needle, the central part of the "intentional threads" that tie it to the given objects. What we move is never our objective body, but our phenomenal body, and there is no mystery to that, since it was already our body, as power of this or that region of the world, that stretched out toward the objects to be taken and perceived them. . . . The work table, the scissors, the pieces of leather offer themselves to the subject as poles of action, they define by their combined values a certain situation, an open situation that calls for a certain mode of resolution, a certain labor.[32]

This describes well the disturbance made by the idea of oral literature: the substitution, for texts, of a "phenomenal body" for the "objective body" we are used to consider. What is the *Iliad*? Well, it is a poem in twenty-four books or chants, allegedly written by Homer about whom almost nothing is known, and it tells about one episode in the Trojan War, starting from the quarrel of Achilles and Agamemnon; it consists of over fifteen thousand hexameter verses. . . . But what is the *Iliad* as "an open situation that calls for a certain mode of resolution, a certain labor"? For guidance on that question, we should ask Paulhan, Meillet, Jousse, Parry, Gesemann, Jakobson-Bogatyrev, Honko, and others who have dealt with attested works as gateways to their limitless potential realizations, as if the existing texts were by-products. Masterpieces, it may be, but by-products still. "Things are said . . . which make one stare."[33]

Merleau-Ponty also describes a man who, injured in the head in the First World War, is unable to form projects.

> If one asks him to trace a square or a circle in the air, he first "finds" his arm, then projects his hand forwards, as a non-disabled person does to detect a wall in the dark, finally he tries out different movements following a straight line and following different curves, and if one of these movements proves to be circular, he completes it promptly. . . . Visibly this patient possesses his own body only as an amorphous mass in which only effective movement introduces

divisions and articulations. He depends on his body to carry out the movement like an orator who cannot speak a word without relying on a text written in advance. . . . What he lacks is not motricity, nor is it thought, and we are led to recognize between movement as a third-person process and thought as a representation of movement, an anticipation or a capture of the result accomplished by the body itself as motor capacity, a "motor project" (*Bewegungsentwurf*), a "motor intentionality" without which the order [to move] remains a dead letter. . . .

Patients less seriously affected than Schn. [the previously mentioned case] are known, who perceive shapes, distances and objects themselves, but are able neither to trace on those objects the directions suitable for action, nor to distribute them according to some rule, nor in general give the spectacle of space the anthropological determinations that build up the landscape of our action. . . . The world exists for them only as a ready-made or frozen world, while for a person without this disability, projects polarize the world and as if by magic plant all over it a thousand signposts to guide action.[34]

Though planted in 1945, Merleau-Ponty's signpost already guides our inquiry in the direction of the theories of embodied cognition that will perhaps enrich our accounts of memory and oral tradition. We have seen in the body (including the brain) and its limitations some of the reasons for the organization of texts around cues that permit near-automatic recall, efficient data compression, and transfer from mind to mind. A reciter without those limitations, a robot or an angel, would not have to organize the "mental text" in that way. An oral text is an interplay between a mental text (largely made up of recall of heard recitations) and a projected text to be realized. This latter is enacted—made to come about—by its reciters in a particular way. The theorists of enactivity hold that "cognition is not the representation of a pregiven world by a pregiven mind but is rather the enactment of a world and a mind on the basis of a history of the variety of actions that a being in the world performs."[35] Analogously, the enactment of an *Iliad* and a Homer might best be imagined "on the basis of a history of the variety of actions that a being in the world performs." The epic—or the consciousness of it in the reciter's mind—could be described as an autopoietic machine.

An autopoietic machine is a machine organized (defined as a unity) as a network of processes and production (transformation and destruction) of components which: (i) through their interactions and transformations continuously regenerate and realize the network of processes (relations) that produced them; and (ii) constitute it (the machine) as a concrete unity in space in which they (the components) exist by specifying the topological domain of its realization as such a network.[36]

We might recall that "concrete unity in [imaginary] space" known as Homeric poetry, which its reciters "constituted," by "patching" it (οἶον ἀκουμένους), "linking and stitching" its parts "closer to one another" (εἰρμῷ τινὶ καὶ ῥάφῃ παραπλήσιον ποίειν) until it composed "a single, continuous, unified, and harmonious body" (ἕν . . . τι σῶμα συνεχὲς διόλου καὶ εὐάρμοστον).[37] Indeed Jerome McGann has maintained that texts are "autopoietic mechanisms operating as self-generating feedback systems that cannot be separated from those who manipulate and use them."[38] That will be welcome news to any author who has ever left a manuscript in a taxicab. But when poems are inscribed on their hearers, poet and poem can be described as a nested pair of autopoietic machines—the poet a living organism in constant exchange with an environment, the poem a shared imagination in constant interchange with its own features and with its hearers. We have no way of knowing in detail how the bards and rhapsodes of other times "continuously regenerated and realized" the poems they heard and reperformed, but surely recognizing and unconsciously adding markers of coherence was one way for a particular song to "define itself as a unity" in the sea of epic poetry and to "constitute itself as a unity in space," whether it was heard in a performing space or written down on a two-dimensional surface. Forgetting a marker, even if it might temporarily endanger the coherence of the "mental text," could well result in a new "transformation of components" intended to be less vulnerable to decay, for a central task of autopoiesis is maintaining boundaries between inside and outside. The embodiment of the performer and the embodiment of the poem are similar processes, invading and modifying one another over the history of the poem's transmission through many poets.

Collective poetry, self-organizing texts, techniques of the reciting body, translations of gesture, this media theory that is not one: the topic of "oral literature" rolled up into itself a great many disciplines and agendas (indeed, some extraordinarily improbable agendas) in the first part of the European twentieth century. The subset of that theorizing that made it across the Atlantic, to be decanted by Parry, Lord, McLuhan, and Ong into a thousand dissertations, was diluted in anticipation of its receivability by the profession of letters in America. But the power of the idea of oral literature should not dissipate in antiquarianism, in salvage ethnography, or even in "secondary orality" with its utopias of electronic participation. Oral literature, throughout the twentieth century, served as an experimental base for both artistic and theoretical avant-gardes, proposing the disappearance of the author, modular composition, automatic writing, self-reference, the iterative, interactive, and ergodic modes of performance; dissolving the unity of the work and extending its moment of creation over many centuries; bringing matter and form together in new and nonhierarchical relations. The unities that oral composition and transmission perturb are precisely those on which the characteristic media of modernity depend: work, word, author, period, nation, doctrine.

It is as if nothing is written.

Acknowledgments

I have the pleasant task of remembering the people who inspired, stimulated, or inflected this project. Anne and François Desgrées du Loû, and their children and grandchildren, have been with me in one way or another throughout. Peter Burian, with a remark dropped to his first-year Greek class when we were struggling through *Iliad* 1, first made me think about Parry and epithets. Inez Hedges was my first guide through the early twentieth-century avant-garde. Jay Geller introduced me to Jousse and Leroi-Gourhan. One tentacular idea came to me while I stood on a ladder painting a ceiling at the house of Manos and Clare Anglias in Athens. Helen Nagge and Mel Chin, Anne-Françoise and Xavier Franque, and Catherine and Stanislas Maillard have been the kind of steadfast friends for whom one writes. My most obvious intellectual debt is to Jacques Derrida, to whose early work this book offers a hain-teny response. Kang-i Sun Chang, Paul de Man, Geoffrey Hartman, Giuseppe Mazzotta, Rulon Wells, and Tom Cole were patient with a wayward student. Roger Blood was exceptionally attentive to holes in my education. I gleaned ideas and gained perspective through memorable conversations with Stephen Anderson, Bernard Bate, Yves Beaupérin, Timothy Billings, Chloe Blackshear, Anna Maria Busse Berger, Robert Brain, E. Bruce Brooks, Chris Bush, Alice Conklin, Joan Connolly, Pascal Cordereix, Jonathan Culler, Marcel Detienne, Jacob Edmond, Jason Escalante, Elisabeth and Jean-Ghislain d'Eudeville, Lazar Fleishman, Alexandre Gefen, John Goldsmith, the late Jack Goody, Rémy Guérinel, Sepp Gumbrecht, David Gutherz, Tom Hare, Barbara Harshav, Eric Hayot, Joshua Katz, Dominick LaCapra,

Dominique Maillard, Boris Maslov, Joe Metz, Carrie Noland, Marjorie Perloff, Elizabeth Povinelli, Philip Rubin, Mark Saler, Pierre Scheffer, S.J., Jeffrey Schnapp, Richard Sieburth, Edgard Sienaert, Michael Silverstein, Pierre-Yves Testenoire, Galin Tihanov, Q. S. Tong, Katie Trumpener, Steve Yao, Grant Wiedenfeld, Zhang Longxi—many of whom have had the kindness to answer follow-up queries as well. I remember departed friends of the cause: Jean-Eudes de la Baume, Richard Maxwell, Michael Toussaint Stowers, Benjamin Harshav, and Tony Yu. I am grateful to my Stanford history-of-science posse (Tim Lenoir, Bob Batchelor, Steven Meyer, Roger Hart) and my New Haven mind-and-language gang (Elise Snyder, Michael Holquist, Mel Woody, James Phillips, Marc Rubenstein, and Joe Errington) for the core goods of intellectual companionship: attention, corrections, and pizza. Without the Bibliothèque Mazarine and the libraries of Stanford, Yale, and Chicago the project would have been unthinkable. I spoke to various groups on parts of this project, and each time the audiences taught me much. So I gladly thank the Department of Folklore and Ethnomusicology, UCLA; St. Joseph of Arimathea Theological Seminary, Berkeley; the Department of Comparative Literature at Yale; the Mind, Brain, Culture and Consciousness Workshop at Yale; the Université de Lyon-3; the Association Marcel Jousse; the School of Theory and Criticism at Cornell; Berkeley College, Yale (and its then Master, John Rogers), the University of Utah, Duke University, the Franke Center and the Musicology Workshop at the University of Chicago, and the De Carle Lecturership at the University of Otago, organized by Jacob Edmond. I thank the deans who have helped this book along by approving and funding leave time: Pauline Yu (UCLA), Sharon Long (Stanford), Jon Butler (Yale), and Martha Roth (Chicago). I did bulimic reading and obsessive outlining during a year at the Stanford Humanities Center: thanks to its director John Bender and his advisory board. Wlad Godzich and Daniel Heller-Roazen, readers for Fordham University Press, were generous with corrections and ideas, and Gregory McNamee, my copy editor, got me to rethink many unclear formulations. A Guggenheim fellowship, though designated for another manuscript, greatly encouraged me to finish this one. A slow-growing project is bound to be interrupted often, and I am thankful for five interruptions in particular: Juliana, Caleb, René, Constantin, Kirill. The late Helen Tartar's engagement kept the manuscript alive through times when I doubted anyone would want to read it. Paul Farmer has accompanied the book's gestation for thirty years with brotherly concern. As listener, critic, and witness, Olga Solovieva has done more to bring it to completion than anyone who was not actually pressing on the keys.

Notes

Foreword

1. McLuhan, *The Gutenberg Galaxy*, 1–3.
2. Manovich, "What Is Digital Cinema?" 2–3.
3. Derrida, *De la grammatologie*, 88–95.
4. Hayles, *Writing Machines*, 103.
5. Ibid., 40.
6. Ibid., 130.
7. von Kleist, "Über das Verfertigen der Gedanken beim Reden."

Introduction: Weighing Hearsay

1. These are usually more advantageously considered as forms of another issue such as class or gender. My subject here is precisely what blocks easy access to theme—a formal or hermeneutic structure. On the struggle between orality and literacy as a dynamic of culture, sometimes reflected in literary themes or devices, see for example Bakhtin, "The Problem of Speech Genres"; de Certeau, "Pour une nouvelle culture: prendre la parole" and *La prise de parole*, esp. 29–37, 177–182, 262–263; and for an analysis coinciding with de Certeau's terms but arriving at an opposite judgment, Barthes, "L'écriture de l'événement"; Brathwaite, *History of the Voice*; Gates, *The Signifying Monkey*. On the written evocation of oral speech in literary texts, see Eikhenbaum, "The Illusion of *Skaz*" and Bakhtin, *Problems of Dostoevsky's Poetics*. Distinguishing "orality" from "oral literature," Kirshenblatt-Gimblett ascribes a leading role to the phonograph in the formation of "orality" ("Folklore's Crisis," 283, 309–319). Many of the questions to which the present book attempts to provide an answer are trenchantly framed in Gumbrecht, "Rhythm and Meaning."
2. Magoun, "The Oral-Formulaic Character of Anglo-Saxon Narrative Verse," 446–447. The passage reads at greater length: "the characteristic feature of all orally composed poetry is its totally formulaic character. . . . Oral poetry, it may be safely

said, is composed entirely of formulas, large and small, while lettered poetry is never formulaic, though lettered poets occasionally consciously repeat themselves or quote verbatim from other poets." Magoun was an advisor for Albert Lord's 1949 Harvard doctoral thesis, published in 1960 as *The Singer of Tales*. The 1953 article draws on Lord's work with Milman Parry.

3. For a study of the graphic aspects of this literature, see O'Keeffe, *Visible Song* and "The Performing Body on the Oral-Literate Continuum." Finnegan, *Oral Poetry*, 23–24, succinctly dispatches the idea that orality is a cultural phase made obsolete by the arrival of literacy. Honko focuses on the processes linking oral song and printed text: "Text as Process and Practice: The Textualization of Oral Epics," 3–54 in Honko, *Textualization of Oral Epics*.

4. See Parry, "Introduction," *The Making of Homeric Verse*, xxviii, xxxi; Parry, "Čor Huso," ibid., 439–440.

5. The songs collected on these trips are now preserved (in the form of notebooks and phonograph discs) in the Milman Parry Collection of Oral Literature, Widener Library, Harvard University.

6. McLuhan, *The Gutenberg Galaxy*, 3. On McLuhan's debt to Parry and Lord, see Finnegan, *Oral Poetry*, 254–260. On the place of media in Western culture generally, see Guillory, "Genesis of the Media Concept."

7. Havelock, *Preface to Plato*, 47–48, 127, 236.

8. Goody and Watt, "The Consequences of Literacy," 307–308, 310–311.

9. Ong, *Orality and Literacy*, 36–54. For examples of the checklist style of citation, see Farrell, "Early Christian Creeds and Controversies"; Andersen, "Oral Tradition," 31–32; Young, *Jesus Tradition in the Apostolic Fathers*, 70–91. For a critique of Ong's spiritualism (hardly surprising in itself) and ethnocentrism, see Sterne, "The Theology of Sound."

10. E.g., Lévy-Bruhl, *Les fonctions mentales dans les sociétés inférieures* (1910).

11. Fabian, *Time and the Other*, 25 (on Ong, see 118–123); on temporality with reference to Ong, see Sterne, "The Theology of Sound." De Vet, in "Parry and Paris," makes a claim for the determining influence on all Oral Theory of Lévy-Bruhl's ideas about the "primitive mind."

12. Holoka, "Homer's Originality," 257.

13. Finnegan, *Oral Poetry*, 15, 210, 201, 232, 24, 22, 260.

14. See, e.g. Harry Levin, introduction to Lord, *The Singer of Tales*, xiii. For a review of the question, see Zumthor, *La lettre et la voix*, 9–25. Le Blanc, "Littératures orales," finds in "oral literature" the means of a thoroughgoing questioning of the categories of "literature."

15. On the distinction between "impossibilia" and "oxymoronics," see Eco, *Foucault's Pendulum*, 74–75. Zumthor, *La lettre et la voix*, 216–217, likens the fixity of oral formula to Derrida's category of "arche-writing." But for a skeptical appreciation of Zumthor's gesture, see O'Keeffe, "The Performing Body on the Oral-Literate Continuum," 51–52.

16. Caesar, *Commentarii de Bello Gallico*, 2:160–164.

17. Plato, *Phaedrus*, 275 a–276 a. On the implications of Plato's rejection of writing, see Derrida, *De la grammatologie*, 55, 58–59, 73, and "La pharmacie de Platon," *La dissémination*, 71–198.

18. On this passage, see Dumézil, "La tradition druidique et l'écriture: Le vivant et

le mort"; and on the Greek and Roman reception of Celtic culture generally, see Momigliano, *Alien Wisdom*, 50–73.

19. Flavius Josephus, *The Genuine Works*, trans. Whiston, 4:364–365; *In Apionem* I.2, in *Flavii Iosephi Iudaei opera omnia*, 5:172.

20. Josephus, *The Genuine Works*, 4:364–365, 367–368; *In Apionem* I.2,8, in *Flavii Iosephi Iudaei opera omnia*, 5:172, 176. On Josephus's persistent references to writing, see Barclay, trans., *Against Apion*, 13–14, and on the argument about historical veracity generally, 15–18, 29–32.

21. Eustathius of Salonica, comments on *Iliad* book 1, in *Commentarii ad Homeri Iliadem pertinentes*, ed. van der Valk, 1:9–11. On Pisistratus's recension, see also Cicero, *De oratore*, 3.34.

22. See Plato, *Phaedrus*, 278 a 3, on the legitimate writing on the soul as writing's virtuous counterpart.

23. A special branch of knowledge grew out of the need to assess the reliability of particular *hadith* and reconcile possible contradictions with scripture. On the debates surrounding the authority of *hadith*, see Musa, *Hadith as Scripture*, esp. 3–14.

24. On the shifting relative places of written and oral testimony in the common law, see Langbein, "Historical Foundations of the Law of Evidence" and Woolf, "The 'Common Voice,'" 39–40. The admissibility of oral tradition in land rights cases involving aboriginal peoples has been much discussed in Canadian and Australian courts, among others. See Miller, *Oral History on Trial*, esp. 3–9.

25. Woolf, "The 'Common Voice,'" 37. On the origin of the term "oral history," see Lepore, "Joe Gould's Teeth."

26. [d'Aubignac], *Conjectures académiques ou Dissertation sur l'Iliade*, 5. Although printed only in 1715, d'Aubignac's speech dates from 1664 and was influential in the circle of the Perrault brothers, instigators of the Quarrel. Though d'Aubignac asserts that his disbelief in Homer is answerable only to pure freedom of inquiry and should not make him suspect of "disloyalty to the Crown or religious infidelity" (6), his iconoclasm served Cardinal Richelieu's nascent absolutist cultural project, according to Fumaroli ("Les abeilles et les araignées," in Lecoq, ed., *La querelle des Anciens et des Modernes*, 80–81).

27. d'Aubignac, *Conjectures académiques*, 98.

28. Ibid., 92.

29. Perrault, *Parallèle des Anciens et des Modernes* (1688), with explicit reference to d'Aubignac's *Conjectures*, in Lecoq, ed., *La querelle*, 376.

30. Perrault, *Histoires ou contes du temps passé*, also known as *Contes de ma mère L'Oye* (1697).

31. Houdar de la Motte, *L'Iliade, Poème, avec un Discours sur Homère*; rebuttal in Gacon, *Homère vengé* and Dacier, *Des causes de la corruption du goût*. Gacon repeatedly ridicules Houdar's "Ode on Criticism" and his pretentions to critical discernment. Believing that there was but a single Homer, Houdar thought that the flaws in his work were failures of taste ("Discours sur Homère," x–xvii). On the contemporary reception of Houdar's preface and translation, the second act of the "Quarrel" begun in 1664, see Lecoq, ed., *La querelle des Anciens et des Modernes*, 450–555.

32. On the critical missions of textual scholarship, see Grafton, *Forgers and Critics*. On the author-centered implications of such Higher Criticism, see Rabau, "Pour une poétique de l'interpolation."

33. Küster, *Historia critica Homeri*, 94–95.

34. Vico, *La scienza nuova*, 580–581. See also Simonsuuri, *Homer's Original Genius*.

35. Vico, *La scienza nuova*, 570.

36. Ibid., 578.

37. Rousseau, *Essai sur l'origine des langues*, 389–390.

38. Vico, *La scienza nuova*, 544–546.

39. Wolf, *Prolegomena ad Homerum*, 28; *Prolegomena to Homer*, 68.

40. Ibid., 30; 69. Bérard's *Un mensonge de la science allemande*, despite its patriotic violence, persuasively demonstrates Wolf's dependence on previous sources, chiefly encyclopedias.

41. On the poems ascribed to Ossian, see Haugen, "Ossian and the Invention of Textual History." On the romantic-regionalist fantasy of the oral poet as related to minority traditions and the collecting of folklore, see Trumpener, *Bardic Nationalism*. Marmier's attention to poetry in dialect goes with an assertion that "oral tradition" conserves memory equally as well as libraries and archives ("Poésies populaires de nos provinces," 1835).

42. On "Wolf's model: Eichhorn," see *Prolegomena to Homer*, 18–26, 85, 141, 146, 227–231; Van Seters, *The Edited Bible*, 191–196.

43. On the early history of the Higher Criticism, see Gibert, *L'invention critique de la Bible*. Simon's *Histoire critique* was at first confiscated and destroyed on the orders of Bossuet, but copies surfaced in Holland and were reprinted in 1680. Samuel Reimarus never published his doubts about the origin and veracity of the Gospels in his lifetime. Lessing initiated a controversy when he made fragments from Reimarus's manuscripts public in 1774–78. See Nisbet, "Introduction," in Lessing, *Philosophical and Theological Writings*, 6–14.

44. On Lessing's "germinal breakthrough," see Farmer, *The Synoptic Problem*, 3–15; Kelber, *The Oral and the Written Gospel*, 77–80.

45. On Lessing's debates with the pastor Johann Melchior Goeze about the authority of oral and written gospels, see Büttgen, "Lessing et la question du prêche," 235–240.

46. Herder, *Von Gottes Sohn*, 209; see also 312, 321–322.

47. Gieseler, *Historisch-kritischer Versuch*, 94–95.

48. Strauss, *Das Leben Jesu für das deutsche Volk*, 87. Strauss attributes this romantic Wolfianism to Gieseler. But Gieseler does not explicitly refer to Wolf in his *Versuch*, and mentions Homer only once.

49. The poems were described on Macpherson's title page as "collected in the Highlands." Hugh Blair, professor of rhetoric at Edinburgh, enlarged on their proposed ultimate origin as having been "a college or order of men, who, cultivating poetry throughout a long series of ages, had their imaginations continually employed on the ideas of heroism; who had all the poems and panegyrics, which were composed by their predecessors, handed down to them with care; who rivalled and endeavored to outstrip those who had gone before them" (Blair, "A Critical Dissertation on the Poems of Ossian" [1763], in Macpherson, *The Poems of Ossian*, 99).

50. Anonymous, "Essay on American Language and Literature," *The North-American Review and Miscellaneous Journal* 1 (1815), 307–314.

51. Montgomery, *Lectures on Poetry and General Literature*, 270, 275.

52. La Landelle, *L'âme et l'ombre d'un navire*, 2.

53. Luzel, "Contes et récits populaires des Bretons Armoricains"; Sébillot, *Littérature*

orale de la Haute-Bretagne. Eliade's account ("Littérature orale") ranges from folktales to mythological archetypes.

54. Chasles, *Michel de Cervantes*, 341.

1. Poetry Without Poems or Poets

1. Paulhan, *Les hain-teny merinas, poésies populaires malgaches*, 8–9. A thoroughly revised version of this work appeared as *Les hain-tenys* (1939); this second version is reprinted in Paulhan's 2009 *Oeuvres complètes*, 2: 131–166 (where the present passage appears on 144–145). For a reassessment of Paulhan's studies of hain-teny by a Malagasy specialist, see Faublée, "Jean Paulhan malgachisant." On Paulhan's "allegories of ethnography," see Syrotinski, *Defying Gravity*, 25–46. See also Paulhan, *Jean Paulhan et Madagascar*. Paulhan later became a central figure in French literary life as critic, novelist, and editor (1925–40) of the *Nouvelle Revue Française*. See Paulhan, *On Poetry and Politics*; Milne, *The Extreme In-Between*; Culbert, *Paralyses*, 174–186; Coustille, "Pour une soutenance de thèse de Jean Paulhan."

2. Paulhan, *Les hain-teny merinas*, 2; *Oeuvres complètes* (2009), 2: 15.

3. Unsure about the field to which his research belonged, Paulhan in 1912 and again in 1936 tried to enlist the anthropologist Lucien Lévy-Bruhl and the linguist Antoine Meillet as advisors for a thesis that in the end would never be written. See Charles Coustille, "Pour une soutenance de thèse de Jean Paulhan," par. 10, and Paulhan, *Jean Paulhan et Madagascar*, 254–265.

4. Paulhan, *Les hain-teny merinas*, 9–10, 59.

5. Ibid., 10–12.

6. Ibid., 15. Compare Bakhtin on the "ultimate dialogicality" of Dostoevsky's novels ("Everything . . . tends toward dialogue. . . . A single voice ends nothing and resolves nothing. Two voices is the minimum for life": *Problems of Dostoevsky's Poetics*, 18, 252) and Jean-Luc Nancy on the condition of "literary communism": "The text interrupts itself at the point where it shares itself out—at every moment, to you, from him or her to you, to me, to them. . . . Literature inscribes being-in-common, being for others and through others" (*The Inoperative Community*, 64, 65–66).

7. Taylor, "Irreducibly Social Goods," in *Philosophical Arguments*, 127–145.

8. Paulhan, *Les hain-teny merinas*, 15.

9. Ibid., 61–62; see also 71.

10. Ibid., 41–50; *Oeuvres complètes* (2009), 2: 153. The 1939 version drops the paragraph on the "two or three hundred rhythmic phrases," perhaps because Paulhan there wishes to emphasize the strategic occasion rather than the permanent vocabulary. On proverbs, see further Paulhan, "L'expérience du proverbe." Patterned cloths, "*lamba hoany*"—bearing a regular repetition of proverbial motifs as if in visual emulation of the recurrent proverbs in *hain-teny* poetry—function as communicative devices in Madagascar (Stewart, *Patternalia*, 116).

11. Ibid., 50; *Oeuvres complètes* (2009), 2:158–159.

12. Ibid., 50–53.

13. Ibid., 41–48.

14. Baudelaire, third "Projet de préface pour les *Fleurs du Mal*," *Oeuvres complètes*, 1:183.

15. Mallarmé, "Crise de vers," *Oeuvres complètes*, 368. Compare Robert de Souza, defending and summing up the accomplishments of Symbolisme in 1906: "A poem has

to be constructed differently from a speech, from a dissertation or a story. But until the Symbolists came, this had never happened." (*Où nous en sommes*, 41.)

16. Quoted, de Souza, *Où nous en sommes*, 18–19 (the anti-Symbolist sottisier extends from 8 to 29). For another defense, see Smith, *Mallarmé's Children*. Maurice Leblond, son-in-law of Emile Zola, founded a literary movement called "naturism" (to be confused with neither Naturalism nor nudism); Paul Souchon, author of *Élévations poétiques* (1898) and *Nouvelles Élévations poétiques* (1901), recycled his warning of a "starless night" in the preface to his tragedy in verse, *Phyllis* (1905).

17. Valéry, "Lettre sur Mallarmé" (1927), *Oeuvres*, 1:638.

18. Valéry, "Je disais quelquefois à Stéphane Mallarmé" (1931), *Oeuvres*, 1:646.

19. Valéry, "Je disais quelquefois," *Oeuvres*, 1:658. In similar fashion, Valéry's fictional hero Monsieur Teste had "killed the marionette" in himself: He did not gesticulate ("La Soirée avec M. Teste," *Oeuvres*, 2:17).

20. "In the end, I was persuaded by some Daimôn or other to consider the self-awareness of my thinking as superior to its productions, and the meditated act as superior to spontaneous productions (even very beautiful ones) or to effects of chance (even lucky chance)—in a word, as superior to anything that can be credited to automatism." Valéry, "Réponse" (1932), in *Oeuvres*, 2:1605.

21. See Marjorie Perloff, "Screening the Page/Paging the Screen: Digital Poetics and the Differential Text."

22. See Hayot and Wesp, "Style, Strategy and Mimesis in Ergodic Literature."

23. Paulhan, "D'un langage sacré," *Jean Paulhan et Madagascar*, 312. This essay also appears in Hollier, *Le Collège de Sociologie*, 694–728; and, translated by Betsy Wing, as "Sacred Language," in Hollier, *The College of Sociology 1937–39*, 306–321.

24. Paulhan, "L'expérience du proverbe," *Oeuvres complètes*, 2:106; "D'un langage sacré," 321. Gumbrecht could have advised him that "a constitutive tension exists between the phenomenon of rhythm and the dimension of meaning" ("Rhythm and Meaning," 171).

25. See Hardin, "The Tragedy of the Commons."

26. For a comical instance of proverbial argument creeping into written language, see Coustille, "Pour une soutenance," par. 20.

27. Cf. Perloff, *Unoriginal Genius*.

28. Dugas-Montbel, *Observations sur l'Odyssée d'Homère*, 133–134.

29. Renan, *L'avenir de la science*, 94–95.

30. The second edition (1929) is identical with the first. Granet refers to Paulhan's *Les hain-teny* on 30, 232, 268–299, and implicitly at many other points; the word "*joute*" (joust), for example, appears no less than sixty-six times. In turn, Paulhan cited Granet's use of his text in seeking the support of potential thesis advisors in 1936 (Jacqueline Paulhan, *Jean Paulhan et Madagascar*, 258–259).

31. On the use of symmetry and opposition to build semantic wholes, see Saussy, "Rhyme, Repetition and Exchange in the *Book of Odes*." For a linguistic analysis of *Shijing* meter and grammar in terms of optimality constraints, see Zuo Yan, "Classical Chinese Verse-Grammar," 17–24.

32. *Shijing* 12 ("Quechao"). Compare the translation by Karlgren, *The Book of Odes*, 7.

33. On the collection's traditional exegesis and the twentieth-century emergence of rival interpretations, see Van Zoeren, *Poetry and Personality*, and Saussy, *The Problem of a Chinese Aesthetic*. For a discussion of the poems as exhibiting oral characteristics, see Kern, "*Shi jing* Songs as Performance Texts."

34. Granet, *Fêtes et chansons*, 232–233 (emphasis in original)

35. Ibid., 89–91. In the original, only the words "sentimental themes" and "rustic themes" are emphasized; the other italics are mine, calling attention to the privative characterization and its implied opposite. See also Roger Bastide's use of Granet's work in reconstructing one stage of a developmental model of art: coming after a primitive stage in which the crowd sings as one, the poetic duel initiates individual self-consciousness, "corresponding to stages of sociological evolution" (*Art et société* [1977 (1945)], 72–74).

36. Ibid., 216.

37. Ibid., 93, 92.

38. Ibid., 235.

39. On collective representations, see Durkheim and Mauss, "De quelques formes primitives de classification." Analogously, in the opening pages of *Formes élémentaires de la vie religieuse*, Granet's teacher Durkheim had dismissed investigation into the content of individual beliefs as profitless and advocated research into the shared, outward testimony of ritual and narrative, which alone furnish manageable objects for the sociologist's inquiry (25). Perhaps because both sociology and Chinese studies were disciplines poorly anchored in the French intellectual world of the time, it seemed advantageous to enlist the two in a mutually supportive argument.

40. Granet, *Fêtes et chansons*, 228–229, 94.

41. Ibid., 93.

42. Durkheim, *Formes élémentaires*, 26 n 1.

43. Granet's later works for a wide public, *La pensée chinoise* and *La civilisation chinoise*, return to the theme of the rhythmical origin of concepts. On the "total social fact," see Marcel Mauss, "Essai sur le don" (1924), *Sociologie et anthropologie*, 147–148, 275–276.

44. Marcel Jousse's major publications are *Le style oral* (1925) and a posthumous trilogy assembled from his articles and lecture notes: *L'anthropologie du geste, La manducation de la parole* and *Le parlant, la parole et le souffle* (1974–78; republished in one volume, 2008). The stenographic transcripts of Jousse's lectures (1931–1957) have been digitized by the Association Marcel Jousse. On Jousse's life and influence, see Lefèvre, *Marcel Jousse: Une nouvelle psychologie du language*; Benjamin, "Probleme der Sprachsoziologie"; Ong, *Presence of the Word*, 147–148; Baron, *Mémoire vivante* (with an ample bibliography of contemporary notices, 70–72); Certeau, "Une anthropologie du geste"; Stephen Heath, "Ambiviolences," 55–57; Meschonnic, *Critique du rythme*, 647–713; Sienaert, "Marcel Jousse: The Oral Style and the Anthropology of Gesture"; Guérinel, "Déchiffrer l'énigme Marcel Jousse"; Beck, "La poésie du geste"; and the special numbers "Marcel Jousse: Un' estetica fisiologica," *Il Cannochiale* 1–3 (2005) and "Marcel Jousse: Pour une anthropologie autre," *Nunc* 25 (2011). On Paulhan and Jousse, see Baillaud, "Sur la manducation de la parole." *Le Style oral* was reviewed enthusiastically by Jousse's superior in the Company of Jesus, Léonce de Grandmaison, and with some reservations by Meillet. Ong (*Orality and Literacy*, 20) and Illich (*In the Vineyard of the Text*, 61) credit Jousse, erroneously, with years of residence in the Middle East. Foley (*Theory of Oral Composition*, 14) contradicts Adam Parry in his eagerness to show Milman Parry innocent of Jousse's influence.

45. Jousse, *Etudes de psychologie linguistique: Le style oral rythmique et mnémotechnique chez les verbo-moteurs*, 108, citing Paulhan, *Les hain-teny merinas*, 52–53.

I replace Jousse's straight brackets, indicating his alterations to the source text, with curly brackets. Another restatement of the Paulhan passage occurs in Jousse, "Les lois psycho-physiologiques du style oral" (1931), 8, but there in the mode of free paraphrase, as if to indicate that Jousse has thoroughly assimilated it.

46. Jousse, *Le style oral*, 112. Internal citations refer to Letourneau, *L'évolution littéraire dans les diverses races humaines* (Paris, 1894) and Jastrow, *A Dictionary of the Targumim* (London and New York, 1903). The chapters of the Q'uran are known as *sûras*, etymologically "sequences"; Jousse connects *sûra* to the cognate Hebrew word *shir*, meaning "song" (as in the Song of Songs, *Shir haShirim*), and, more punningly, to Hebrew *keshir* (link, tie). (I thank Avner Greif for clarifying these details.) The argument is that "link" is the root notion of both "song" and "sequence."

47. Jousse, *Le style oral*, 36, citing Ernest Leroy, *Le langage: Essai sur la psychologie normale et pathologique de cette fonction* (Paris: Alcan, 1905), who in turn cites Garrick Mallery, *Sign Language Among North American Indians*, 323 (here restored from the original).

48. See the statutes of the Society (*Mémoires de la Société de Linguistique de Paris* 1 [1868], 111), and Ferdinand de Saussure, *Cours de linguistique générale*, ed. Tullio de Mauro, 417.

49. Jousse, *Le style oral*, 13, paraphrasing Bos (i.e., Marie Boeuf), "Contribution à la théorie psychologique du temps," in Janet, *IVe Congrès international de Psychologie tenu à Paris en 1900* (Paris: Alcan, 1901), 290. On the reflex arc and its importance for nineteenth-century psychology, see Menand, *The Metaphysical Club*, 323–340.

50. "'Primitives'—a word that must disappear from the vocabulary of every scholar" (Marcel Mauss in 1922, cited in Jousse, *Le style oral*, 18). The word is more easily banished than its connotations, especially temporal. On the ethical and epistemological issues surrounding the term, see Lévi-Strauss, "La notion d'archaïsme en ethnologie," *Anthropologie structurale*, 113–132, Fabian, *Time and the Other*, and Li, *The Neo-Primitivist Turn*.

51. Jousse, *Le style oral*, 8–9, citing and retouching Émile Baudin, *Psychologie* (Paris: Gigord, 1919).

52. Ibid., 4, citing and retouching Georges-Henri Luquet, *Idées générales de psychologie* (Paris: Alcan, 1906), and Théodule Ribot, *La vie inconsciente et les mouvements* (Paris: Alcan, 1914).

53. Ibid., 12.

54. Ibid., citing and retouching Pierre Janet, "La tension psychologique et ses oscillations," in Dumas, *Traité de psychologie* (Paris: 1923), 1:919. The model of the body with which Jousse engages is described in chapter 4.

55. Ibid., 23. The terminology of "images" and "motor images" is common to Ribot, Charcot, Binet and their followers, and already controversial by 1900. See Binet, *La psychologie du raisonnement*, 27–31; Bergson, *Matière et mémoire*, 12–17; van Biervliet, "Images sensitives et images motrices."

56. Bergson, *Matière et mémoire*, 89.

57. Jousse, *Le style oral*, 19, citing and retouching A. Antheaume and G. Dromard, *Poésie et folie: Essai de psychologie et de critique* (Paris: Doin, 1908).

58. Ibid., 18–19, citing Ernest Renan, *Vie de Jésus*, and Jérôme and Jean Tharaud, "L'an prochain à Jérusalem," 771. The Tharauds' travel report is remarkable for its expression of contempt toward oriental Jews and European Zionists, an estimate here reversed, though on primitivist grounds.

59. Ibid., 107, with reference to Baudin, *Psychologie*.

60. Ibid., 114.

61. On conflicting models of memory elicited by this question, see Leys, *Trauma*.

62. Jousse's fascination with the body and with the specificity, the *auto*-matism, of the body's laws recalls Vico. Compare Vico on "the poverty of the human mind, which, lying sunken and buried in the body, is naturally inclined to perceive bodily things and must apply extraordinary effort and care if it is understand itself—like the bodily eye which sees all objects outside itself and needs a mirror to see itself" (*La scienza nuova*, 232).

63. Jousse, *Le style oral*, 204.

64. Ibid., 100, citing O. F. Cook, "The Biological Evolution of Language," *The Monist* 14 (1904): 481–491.

65. See Meschonnic, *Critique du rythme*, 692, 695, 699.

66. Jousse, *Le style oral*, 142, 119.

67. Both "cliché" and "stereotype" derive from an eighteenth-century technique, the predecessor of Linotype and other hot-metal typecasting processes, in which a wax or plaster mold was made of a page of set type and from this a single lead plate was cast. This plate is the *cliché* (see *Trésor de la langue française*, s.v. *stéréotype*). For the extended meaning, see Rémy de Gourmont, *Esthétique de la langue française*: "We must here distinguish the *cliché* from the *commonplace* [*lieu commun*]. In my usage, at least, *cliché* represents the utter materiality of the phrase; commonplace, on the other hand, indicates the banality of the idea. The standard for the *cliché* is the proverb, stiff and unchangeable; the *commonplace* takes as many forms as language offers combinations of words to express a stupidity or a platitude" (cited, *Trésor de la langue française*, s.v. *cliché*).

68. Jousse, *Le style oral*, 77, citing abbé Maurice Landrieux, *Aux pays du Christ: Études bibliques en Egypte et en Palestine* (Paris: Maison de la Bonne Presse, 1895), 400, 437–438. Again, Jousse cites a supercilious traveler only to hold up the behavior of the "natives" as meaningful and appropriate.

69. Jousse, *Le style oral*, 63, citing Lucien Arréat, *La mémoire et l'imagination (peintres, musiciens, poètes et orateurs)* (Paris: Alcan, 1895).

70. Ibid., 113, citing van Gennep, *La question d'Homère*, 51–52. Van Gennep relied on secondary sources for his characterization of the "illiterate" (p. 53) guslars.

71. Ibid., 148.

72. Ibid., 67, 66, citing Denis Buzy, *Introduction aux paraboles évangéliques* (Paris: Gabalda, 1912) and Rodrigues, *Les origines du sermon sur la montagne* (1868), 11.
Rodrigues's preface is significant. He addresses it to the French Minister of Education, who had delivered a speech in praise of the "Christian morality" of the Sermon on the Mount. Rodrigues answers him with a book showing that the vocabulary, ideas, and structure of the Sermon are entirely Jewish. His preface quotes Salomon Munk, the editor of Maimonides, as saying: "Some people are surprised that the Sermon on the Mount had so little effect. But it was current in the streets of Jerusalem long before it was pronounced. Nothing would be easier than to make it over again with materials of a previous era." This Rodrigues sets out to do.

73. Meillet, *Les origines indo-européennes des mètres grecs* (1923), 61. It is also worth noting that for Meillet Homeric epic was no folk poetry, but a "learned poetry, handled by specialists . . . a poetry of professionals, made with memorized formulas" (ibid.). The hexameter meter itself fits awkwardly with the prosody of the Greek language

and restricts the choice and placement of words; Meillet surmised it must have had a foreign origin, probably in the pre-Hellenic endogenous culture of the Aegean. On Meillet's studies in comparative metrics, see Bader, "Meillet et la poésie indo-européenne."

74. Platt, review of Meillet, in *Classical Review* 38 (1924): 22, cited in Milman Parry, "The Traditional Epithet in Homer" (1928), in Parry, *The Making of Homeric Verse* (henceforth cited as "Traditional Epithet" and *Making*), 9.

75. Antoine Meillet, review of Jean Paulhan, *Les hain-tenys*, ccclvi.

76. Jousse, *Cours oraux*, Hautes études, 13 December 1944, cited in Testenoire, ms 5. Jousse was led to Paulhan's 1913 book by the psychologist Pierre Janet (*Le style oral*, 79); presumably Janet knew of it through his colleague Frédéric Paulhan, Jean Paulhan's father. On Jousse's relations with Janet, see Guérinel, "Marcel Jousse entre Pierre Janet et Joseph Morlaas" and Lefèvre, "La psychologie expérimentale."

77. Testenoire, ms 5.

78. On Puech's career as editor of the Greek lyric poets and historian of the transition from classical to early Christian literature, see Virolleaud, "Notice." On the advisor's influence, see Lamberterie, "Milman Parry et Antoine Meillet."

79. Reviewed by Meillet in *Bulletin de la Société de Linguistique de Paris* 26 (1925): 5. Meillet's review is guarded. "The merit of M. Jousse's book is to have presented language as a set of bundled movements following rhythms. . . . M. Jousse's book raises the question rather than resolving it. . . . It opens a series of investigations that will enable us to see more realistically than before the rhythmic element of language."

80. See, e.g., Jousse, *Le Style oral*, 113, 148. On Parry's turn to South-Slavic oral epic, see Parry, "Čor Huso," in *Making*, 439–441; Adam Parry, "Introduction," *Making*, xxii–xxiv. On Murko's work, see Murko, "The Singers and their Epic Songs," and Tate, "Matija Murko, Wilhelm Radloff, and Oral Epic Studies."

81. Mentioning Düntzer, Meillet, van Gennep, Murko, and Jousse as predecessors, Adam Parry says that "[Milman] Parry's achievement was to see the connection between [their] disparate contentions and observations" ("Introduction," *Making*, xxii). Here it is argued that the observations were not so "disparate," but were already drawn into a tight theoretical network. On Parry's absorption of the linguistics and anthropology of his time, see also de Vet, "Parry in Paris." But de Vet's chronology and intellectual history are often imprecise.

82. Arnold, "On Translating Homer," 9–10.

83. Parry, "Traditional Epithet," 120–122.

84. Henri Estienne (Henricus Stephanus), *Certamen Homeri et Hesiodi*, 75; Houdar de la Motte, *l'Iliade, Poème, avec un discours sur Homère*, xv, lxv.

85. For a survey, see Parry, "Traditional Epithet," 119–131.

86. Ibid., 127, 132.

87. Ibid., 151–152.

88. Parry, "Homer and Homeric Style," in *Making*, 266.

89. Parry, "Traditional Epithet," 137.

90. Ibid., 19.

91. Ibid., 51.

92. Ibid., 13.

93. Ibid., 19.

94. Ibid., 104.

95. "Extension" and "simplicité" are the terms of Parry's original publication in

French (see, for example, ibid., 7); Parry later translates his own expressions as "the length and thrift of a system of formulas" in "Homer and Homeric Style," in *Making*, 276. "Extension" for the number of alternatives available to the poet and "economy" for the absence of redundant alternatives (i.e., formulas equivalent in both meaning and meter) have been the terms in use since Hainsworth, *The Flexibility of the Homeric Formula* (1968). For a careful study of the claims made by Parry and his followers, see Dominicy, "La formule chez Milman Parry." For a reading of Parry that denies extension and economy and hopes thereby to save the creativity of the individual poet from determinism, see Shive, *Naming Achilles*.

96. Parry, "Traditional Epithet," 94.

97. Saussure, *Course in General Linguistics*, 117. For further analogies between Saussurian linguistics and Parry's account of formula, see de Vet, "Parry and Paris," and Saussy, "Writing in the *Odyssey*," 308–310.

98. Parry, "The Traditional Epithet," 23, 133, 144–145; "Homer and Homeric Style," 268–269.

99. Parry, "Homer and Homeric Style," 266; "The Homeric Language as the Language of an Oral Poetry," 325. The articles were first published in 1930 and 1932. The second is largely concerned with features of vocabulary and accidence special to epic—thus "language" in a nonmetaphorical sense—but explains their occurrence by the constraints of meter and recitation. Adam Parry speaks of an "artificial language" of epic (*Making*, xxxiv), echoing Kurt Meister, *Die homerische Kunstsprache* (1921).

100. Parry, "Traditional Epithet," 137; see also *Making*, 268–269. For the converse of Parry's observation, see Martin, *The Language of Heroes*, 196: "To read Achilles' speech properly, we are obliged to reread every scene in the *Iliad* in which any phrase of that speech appears."

101. Parry, abstract prefixed to "Whole Formulaic Verses in Greek and Southslavic Heroic Song" (1933), in *Making*, 376.

102. Bogatyrev and Jakobson, "Die Folklore als eine besondere Form des Schaffens," in *Donum Natalicium Schrijnen*, 904; Jakobson, *Selected Writings*, 4: 6; trans. Manfred Jacobson as "Folklore as a Special Form of Creativity," in Steiner, *The Prague School*, 38 (translation modified). For the definition of "langue," see Saussure, *Course*, 19; *Cours*, 38. On Saussure's engagement with orally transmitted texts, see Testenoire, "Littérature orale et sémiologie saussurienne."

103. Kittredge, "Introduction," xvii. For similar articulations, see Gummere, "Primitive Poetry and the Ballad"; or Sharp, *English Folk-Song: Some Conclusions*, 11: variations "are not corruptions. . . . They are the suggestions of individuals, akin to 'sports' in animal and flower life, which will only be perpetuated if they win the approval of the community. . . . Manifestly, those alterations will alone survive which commend themselves to other singers and narrators and are imitated by them." For a history of the problem, see Donatelli, "'To Hear with Eyes.'"

104. The earliest appearance of the word "structuralism" occurs in a 1929 article by Jakobson ("Romantické všeslovanství—nová slávistika," reprinted in *Selected Writings*, 2: 711). For the first occurrence in French (1933), see Troubetzkoy, "La phonologie actuelle," 233.

105. Saussure, *Cours*, 163–167; *Course*, 117–119.

106. Jakobson would provide a preface to Lord's first volume of *Serbocroatian Heroic Songs* in 1954.

107. Saussure, *Cours*, 23–39; *Course*, 8–20.

108. "Whether the formula belongs to the tradition or whether it is, on the contrary, the poet's creation" is irrelevant to the study of Homeric style; "To try to discover in which formula the use of a given epithet is oldest would be pointless" (Parry, "Traditional Epithet," 14, 87). So-called dialect or archaic forms are adopted for their metrical function and, through this adoption, no longer have any relation to place or era ("The Homeric Language as the Language of an Oral Poetry," 327–328).

109. Later formulaic analysts have retreated somewhat from Parry's functionalism. On turnover among them, see Hainsworth, "Good and Bad Formulae." Nagy prefers to speak of "trends" rather than "constants": for his reasons, see Nagy, *Greek Mythology and Poetics*, 24.

110. Saussure, *Cours*, 223–237; *Course*, 161–176. On the coining of new formulas as a case of analogy, see Parry. "Traditional Epithet," 68–74, 175–178.

111. "The neo-grammarians define sound-change as a purely phonetic process; it affects a phoneme or a type of phonemes either universally or under certain strictly phonetic conditions, and is neither favored nor impeded by the semantic character of the forms which happen to contain the phoneme" (Bloomfield, *Language*, 364). For the neogrammarians "all sound changes, as mechanical processes, take place according to laws that admit no exceptions" (Robins, *A Short History of Linguistics*, 183, citing Osthoff and Brugmann).

112. See Saussure, *Cours*, 166–167, on "the countless instances where alteration of the signifier occasions a conceptual change" (*Course*, 121); and for a refutation of the common thesis that linguistic change is led by *Zeitgeist* or national character, see Saussure, "Notes inédites," 61–66. Nagy's work both reintroduces diachrony and reverses the priority accorded to form over meaning in neogrammarian or Saussurian thinking. Where his and Parry's working hypotheses conflict, Nagy has attempted to show that meter originally derives from theme (and is thus conditioned by it). See for example Nagy, *Greek Mythology and Poetics*, 18–35.

113. Cited in de Mauro, "Notes biographiques et critiques," Saussure, *Cours*, 337 (gendered pronouns as in original).

114. Jakobson, *Selected Writings*, 4: 6–7; trans. Jacobson, 38–39 (translation modified).

115. See for example Meier: "Folksong has only an instantaneous form that vanishes with the moment in which it is sounded, and never a fixed and final form, just as happens with the spoken word": *Balladen*, "Einleitung" (1935), 7. On the contrary pulls on folklore in the directions of *langue* and *parole*, see Bauman, "Conceptions of Folklore in the Development of Literary Semiotics."

116. Bogatyrev and Jakobson's essay has many points in common with Tynjanov's 1927 "On Literary Evolution." These points are further accentuated in Jakobson and Bogatyrev's "On the Boundary Between Studies of Folklore and Literature" (1929) and Tynjanov and Jakobson's "Problems in the Study of Literature and Language" (1928). For these studies, see Matejka and Pomorska, eds., *Readings in Russian Poetics*, 66–78, 91–93, and 79–81. The original publication of "Folklore as a Special Form of Creativity" refers to Viktor Shklovsky's work on story types and narrative structure (911; see Shklovsky, *Theory of Prose*, 15–51); this reference disappears in the 1966 reprinting, no doubt owing to personal conflicts between Shklovsky and Jakobson (see Erlich, *Child of a Turbulent Century*, 132–133).

117. Zumthor, committed to the historical specificity of the act of *parole* in oral po-

etry, dismisses such projects: *Introduction à la poésie orale*, 42. The remedy (attention to the oral poem at the moment of its enunciation) follows on 81–83.

118. Bogatyrev and Jakobson, "Folklore as a Special Form of Creativity," 35; Jakobson, *Selected Writings*, 4:2–3.

119. Bogatyrev and Jakobson, "Folklore as a Special Form of Creativity," 43 (translation modified); Jakobson, *Selected Writings*, 4:12. Of course, Jousse is mistakenly credited with the authorship of the passage he cited from Paulhan.

120. Paulhan, cited in Jousse, *Le style oral*, 108; Jousse, cited in Bogatyrev and Jakobson, "Folklore as a Special Form of Creativity." The 1971 Russian translation of the essay reads at this point: "are handed over to oral tradition without modification," which is closer to Paulhan's meaning and reads more intelligibly, but disregards the article and the plural in "die Modifikationen." (Bogatyrev, *Voprosy teorii narodnogo iskusstva*, 380; I thank Boris Maslov for this reference.)

121. Bogatyrev and Jakobson, "Folklore as a Special Form of Creativity," 35; Jakobson, *Selected Writings*, 4:2–3.

122. Ibid., 36; 4:3.

123. Ibid., 39; 4:7.

124. Ibid. (translation modified); 4:7–8.

125. Ibid., 40; 4:8.

126. Ibid., 40–41; 4:8–9. On Homer as *gesunkenes Kulturgut*, see d'Aubignac, *Conjectures académiques ou Dissertation sur l'Iliade*, 21. On Naumann's life, see Nemec, "Naumann, Hans"; on his conception of folklore, see Klaas-Hinrich Ehlers, "Petr G. Bogatyrev: Leben und Werk," in Bogatyrev, *Funktionale-strukturale Ethnographie in Europa*, 40–43; Dow, "Hans Naumann's *gesunkenes Kulturgut* and *primitive Gemeinschaftskultur*." On Soviet responses to Naumann, see Liberman, "Introduction," in Propp, *Theory and History of Folklore*, xlvi (Liberman finds the description of Naumann as "profascist" merely a "typical accusation" for the time).

127. Bogatyrev and Jakobson, "Folklore as a Special Form of Creativity," 41, 40; Jakobson, *Selected Writings*, 4:8.

128. Ibid., 42; 4:10–11, citing Naumann, *Primitive Gemeinschaftskultur*, 190.

129. Ibid., 42; 4:11.

130. Or consider a secret, encoded in a riddle that is passed down for generations and only solved, let us say, hundreds of years after its creation. In this case (a folklore motif in its own right), the container of the secret must have been memorable and worth repeating, whatever the secret itself may be. Or more specifically: it must be made in such a way that at least one hearer in each generation will remember and repeat it. Even if nothing were written down, would we not say that the secret had been *inscribed* in the other narration?

131. Bogatyrev and Jakobson, "Folklore as a Special Form of Creativity," 39; Jakobson, *Selected Writings*, 4:7. Jakobson's ties to avant-garde artists, poets, and theorists are evoked in his book of recollections, *My Futurist Years*.

132. Dow, "Hans Naumann," 61–65. The rectorship would follow in November 1933. Hints of an aristocratic attitude toward the *Volk* would nonetheless cause Naumann trouble throughout the Nazi years (62–63). For an appreciation of Heidegger's philosophy as a return to ancestral German myth, see Naumann, "Sorge und Bereitschaft: Der Mythos und die Lehre Heideggers," in Denker and Zaborowski, eds., *Heidegger und der National-Sozialismus*, 1:178–193.

2. Writing as (One Form of) Notation

1. Lord, *The Singer of Tales*, 30, 35–36. The epigraph is from Parry's fieldnotes, cited by Albert B. Lord, "Homer, Parry, and Huso," in *Making*, 470.

2. Albert Severyns, review of *L'épithète traditionnelle*, 883.

3. Nagy, *The Best of the Achaeans*, 2.

4. Lord, *The Singer of Tales*, 101, 135.

5. Ibid., 220–221.

6. For examples of repetition-counting, see Parry, *Making*, 301–304; Lord, *The Singer of Tales*, 130, 143; and Fenik, *Typical Battle Scenes in the Iliad*, 4–8, 228–231. For Martin, such inventories merely serve to raise "the hard issue . . . of the meaning of repetition within a formulaic art" (*The Language of Heroes*, 170).

7. "To predicate of a culture that it is oral implies that the oral means of communication are made the defining characteristic of the culture in question. We are used to taking writing as a criterion. . . . [Peoples without writing] should not be considered lacking. . . . Their culture is altogether different and complete in its own terms; it is *oral*. It should be interpreted in terms of what it has, that is, orality, and not in terms of what others have, that is, literacy," begins Øivind Andersen ("Oral Tradition," 20–22), but continues: "In an oral culture words take on meanings in specific contexts. There is *no* accumulated word-history and there are *not* codified definitions to abide by. You *cannot* go beyond the use of a word to its 'lexical meaning.' . . . The collective memory or common consciousness in an oral culture . . . is wholly dependent upon the memory of the members of the culture, with *no* inscriptions or written laws to assist it, *no* documents or archives, let alone annals or historical literature. . . . So there can be *no* real sense of history. . . . As there are *no* testimonies which are independent of the oral tradition . . . so there can be *no* recourse to an indisputable, immutable source for 'how it really was' and—more importantly—*no* need is felt for a definitive version of the past because *no* use can be made of it" (22–24; italics mine).

8. See Parry, "Čor Huso," in *Making*, 448–449, on singers corrupted by literacy and the folklore industry ("such singers have no worth"). Katherine O'Brien O'Keeffe, sets herself the task of understanding the manuscript record of Anglo-Saxon poetry as a mixed medium of orality, writing, and bodily gesture (*Visible Speech*; "The Performing Body on the Oral-Literate Continuum").

9. For the phrase and the contrast with vitalism (described as fascist in tendency), see Benjamin, "The Work of Art in the Age of Mechanical Reproduction" (1935–36), *Illuminations*, 217–252; original in two versions, "Das Kunstwerk im Zeitalter seiner technischen Reproduzierbarkeit," *Gesammelte Schriften*, 1.2:435–508.

10. Parry, "Homer and Homeric Style," 317 (italics added). For other instances of the assertion "formulaic, therefore oral," see Parry, *The Making of Homeric Verse*, 269–272, 278, 314–324, 376–378, 389, 392, 451–454, 462–463. Before his first collecting expedition to Yugoslavia in 1933, Parry seems to have depended on Marcel Jousse for his knowledge of preliterate traditions (*Le style oral* is the source of most of the comparative evidence cited in Parry's papers of 1930 and 1932).

11. Parry, "Homeric Language as the Language of an Oral Poetry," 331. For a searching reexamination of the thesis that oral poets compose only in the moment and do not have a means of storage (which would imply writing), see Casajus, *L'Aède et le troubadour*, 24–35, 73–77.

12. Parry, "Homer and Homeric Style," 324.

13. Ibid., 321–322; see also 377–378.

14. Rousseau, *Essai sur l'origine des langues*, 110–111.

15. Wood, *An Essay on the Original Genius and Writings of Homer*, 259–260. On Wood, see Trumpener, *Bardic Nationalism*, 78, 89, 113.

16. Herder, preface to "Volkslieder, zweiter Teil" (1778), *Werke*, 1:244.

17. Lord, *The Singer of Tales*, 148 (italics mine).

18. For a review of the theories about orality and a contestation of the particular properties said to manifest it, see Finnegan, *Oral Poetry*.

19. Goody and Watt, "The Consequences of Literacy"; Havelock, *Preface to Plato*; Ong, *Orality and Literacy*; Goody, *The Logic of Writing and the Organization of Society* and *The Interface between the Written and the Oral*. For a reexamination of the question, see Haas, *Writing Technology*, 5–16, and on the power of myths about technology, 33–47.

20. On the theological flavor of Lévi-Strauss's fourfold analogy among myth and *bricolage*, philosophy and engineering, see Derrida, "La structure, le signe et le jeu dans le discours des sciences de l'homme," in *L'écriture et la différence*, 418.

21. Herder, "Auszug aus einem Briefwechsel über Ossian und die Lieder alter Völker," in *Werke*, 1:866.

22. Lord, *The Singer of Tales*, 99–101 (by "plastic" I think vinyl is meant). See also Milman Parry, "Čor Huso," *Making*, 462.

23. For a detailed and thoughtful study of the reasons for variation in an orally transmitted ritual text, see Goody and Gandah, *Une récitation du Bagré*. The poem as recited by Benima Dagarti includes the self-commentary: "Bagre / is all one; / nevertheless / the way it's performed / is different . . . . You finish Bagre / and yet you don't" (Goody, *The Myth of the Bagre*, 186–197, 190–191).

24. See particularly Goody, *The Logic of Writing and the Organization of Society*, 127–170; Stock, *The Implications of Literacy*, 34–42, 52–62, 86–87, 522–526. The use of oral versus written genealogies as a means of identifying property rights—a body of case law directly related to colonial administration—is a frequent topic in such discussions: see Goody and Watt, "The Consequences of Literacy," 310–311; Vansina, *Oral Tradition*, 51, 78–81, 152–154.

25. On "multiformity" as a positive trait of oral literature, see for example Zumthor, *Introduction à la poésie orale*, 253–54, and Nagy, *Poetry and Performance*, 39–58, 107–114.

26. Skafte Jensen, *The Homeric Question and the Oral-Formulaic Theory*, 85–86 (italics mine). See also Sealey, "From Phemius to Ion," 330: "In the conditions of oral composition . . . poets boast, not of specific songs which they have composed, but of their skill in composition."

27. Lord, *The Singer Resumes the Tale*, 197–198. Lord's fuller explanation: "In my paradoxical description of such poetry as 'written composition without writing,' I was attempting to emphasize that the prolonged and to a very high degree *premeditated* nature of such composition and the *strenuous* efforts to transmit it to others by *rote* memorization resemble more closely a written than an oral *mentality*. To my mind, such poetry would appear to be 'oral' mainly *in its performance*" (198; italics mine). The reputedly letter-perfect transmission of the Vedas and Upanishads for centuries without writing has often appeared as the notable exception in accounts of oral recitation as improvisation. See, for example, Bogatyrev and Jakobson, "Folklore as a Special Form of Creativity," 45.

28. Parry and Lord, *Serbocroatian Heroic Songs*, 1:338 (italics mine).

29. Ibid., 1:239–240.

30. Ibid., 1:240–241.

31. For an inventory of major differences, see Parry and Lord, *Serbocroatian Heroic Songs*, 1:406, 409–413. In all fairness, between 1934 and 1951 Zogić and his region had been through a war and changes of regime that left little intact of the old social order where epic poetry had had its place and audience. For a similar case, with attention to the technology of recording, see Goody, *Myth, Ritual and the Oral*, 58–61.

32. On the advantages of interpreting with "charity," see Davidson, "On the Very Idea of a Conceptual Scheme," *Inquiries into Truth and Interpretation*, 183–198.

33. Gesemann, "Kompositionsschema und heroisch-epische Stilisierung," in *Studien zur Südslavischen Volksepik*, 300.

34. Radloff, "Samples of Folk Literature from the North Turkic Tribes," 83, 87. On another case of field transcription and its far-reaching effects, see Silverstein, "From Baffin Island to Boasian Induction."

35. Radloff, "Samples of Folk Literature," 83.

36. What counts as writing matters for all sorts of reasons (Derrida, *De la gram-matologie*, 15–20). For a traditional history of writing that begins with "primitive semasiology" and moves through ideographs, syllabaries, consonant notation, and finally "full writing," see Gelb, *A Study of Writing*. For some descriptions as "writing" of systems that have often been classified as "proto-writing," see Boone and Mignuolo, *Writing Without Words*. Severi, *Le principe de la chimère*, extends the province of writing to include a large class of mnemonic objects, often three-dimensional or temporal in nature. Ferraris, *Documentality*, similarly seeks to include gesture, memory, and even thinking in the Derridean category of "arche-writing" (214–223). Barber sees a succession of emphases in the field of ethnography: "While performance theory stressed the emergent moment, 'entextualization' theory focused on the ways in which fluid discourse is fixed, and made available for repetition, recreation or copying. . . . Furthermore, structural properties of the text can encourage repetition and thus, by definition, detachment from a single original context" (*The Anthropology of Texts, Persons and Publics*, 71). The possibility of such repetition and detachment, in my view, betokens writing. See also Honko, "Text as Process and Practice" (*The Textualization of Oral Epics*, 3–54).

37. The procedure described here amounts to a reverse-engineering of concordances such as Schmidt's *Parallel-Homer*, which Parry is known to have used. Parry's own representation of repetition and near-repetition ("formulae" and "calembours") in Homer involves adding underlining and dots to the Greek text, oddly complexifying the look of the page when the effect of formula is to simplify. A hypertextual presenta-tion of the epics is now available through the online *Chicago Homer*, edited by Kahane and Mueller. Another electronic Homer edition, one concentrating on attested varia-tions, is described in Dué and Ebbott, "Digital Criticism." Passanante sees presages of a bit-based Homer in the Renaissance revival of atomistic approaches to both nature and text (*The Lucretian Renaissance*, 120–153, esp. 138).

38. See Arend, *Die typischen Scenen bei Homer*; Fenik, *Typical Battle Scenes in the Iliad*; Reece, *The Stranger's Welcome*.

39. Milman Parry, *The Making of Homeric Verse*, 406, emphasis added. The technique of *auxesis* or *amplification* is recognizable in countless Homeric episodes: for consideration of a few examples, see Edwards, "Topos and Transformation in

Homer"; Griffin, "Homer and Excess"; Létoublon, "Le messager fidèle." Indeed the whole of the *Odyssey* may be an *auxesis* of Athena's request, in the opening scene, that Zeus "bring to mind" the forgotten Odysseus. See Kahane, "The First Word of the *Odyssey*."

40. See Merleau-Ponty, *Phénoménologie de la perception*, 140–141, on "Schn.," the man whose brain injury limited his ability to form "projects" of bodily movement. He was reduced to the "what" of movement, impeded from inhabiting its "how."

41. On "remediation," see Bolter, *Writing Space*.

42. See Busse Berger, "L'invention du temps mesuré" and *Medieval Music and the Art of Memory*. Staff notation facilitated not only the transmission but also (even primarily) the composition of polyphonic music. For examples of musical learning in a milieu combining variously fixed and flexible representations (score notation, witnessed and remembered performances, phonograph records), see Berliner, *Thinking in Jazz*, 95–119, esp. 111–112. Goodman, in *Languages of Art*, seems to rejoice in a counter-intuitive stance when he holds that "a score, whether or not ever used as a guide for a performance, has as a primary function the authoritative identification of a work from performance to performance. . . . From this derive all the requisite theoretic properties of scores and of the notational systems in which they are written" (128; see also 129, 156). As Goodman observes, such a definition requires that many things commonly recognized as scores be "reclassified."

43. Korzybski, *Science and Sanity*, 58. Gregory Bateson was responsible for the wider extension of the slogan: see for example "A Theory of Play and Fantasy," *Steps to an Ecology of Mind*, 177–193.

44. See Lü Buwei [attr.], *Lüshi chunqiu* (The Springs and Autumns of Mr. Lü), 5/11b. On the wider implications of musical scales and harmonies in early China, see Falkenhausen, *Suspended Music*.

45. See Lomax, *Mister Jelly Roll*. Jelly Roll Morton's reminiscences, taken down almost accidentally on shellac disk by Alan Lomax, include a compendium of the piano-playing styles of long-forgotten rivals. See Morton, *The Complete Library of Congress Recordings*.

46. For an articulation of such a model in physical and cultural anthropology, involving material conditions and feedback loops, see Durham, *Coevolution*.

47. See, for example, OuLiPo, *La littérature potentielle*.

48. Hesiod, fr. dub. 357, accepted as testimony to the contest of Hesiod and Homer.

49. "[They] collected the scattered, incomplete fragments of Homer's poetry and put them in suitable order." Küster, *Historia critica Homeri*, 87–88.

50. "[They] chose lines suited to their subject from the whole Homeric corpus and joined them together, making, one might say, centons." Küster, *Historia*, 88. *Cento*: "a garment of several bits or pieces sewed together, a rag-covering, patchwork. . . . The title of a poem made up of various verses of another poem" (Lewis and Short, *A New Latin Dictionary*, s.v.).

51. "In this sense, [the rhapsodes] are called ῥαπτῶν ἐπέων ἀοιδοὶ, i.e., singers of stitched songs; which words the scholiasts explain thus: *The poetry of Homer not having been brought together, it lay scattered here and there and divided into parts, and whenever they rhapsodized it, they gathered it into one with a kind of linking or stitching.*" Küster, *Historia*, 87, paraphrasing a scholion also reproduced in Drachmann, *Scholia vetera*, 3:30.

52. Eustathius, *Commentarii*, 9–11, also cited in Küster, *Historia*, 88 (and cf. Drach-

mann, *Scholia vetera*, 3: 29–30). Perrault, relying on d'Aubignac, says similarly: "The term 'rhapsodies,' which in Greek means a mass of different songs stitched together, could not reasonably have been given to the *Iliad* and *Odyssey* except on this basis [of their being later assemblages]." *Parallèle des Anciens et des Modernes* (1688), in Lecoq, *La querelle des Anciens et des Modernes*, 374.

53. Küster, *Historia*, 88. One such example is Petrus Candidus (also known as Decembrius), *Homerokentra, hoc est centones ex Homero* (Venice, 1502). For recent editions and reevaluations of centon collections, see Rey, *Centons homériques*; Usher, *Homeric Stitchings*; Usher, *Homerocentones*.

54. Küster, *Historia*, 89.

55. Not that literary ownership in the modern sense of copyright is necessarily at stake here; questions of authority and legitimacy based on a firmly bounded text can just as well issue in the proud denial of one's authorial role, as in Tertullian:

> I am the apostles' heir. . . . As they ordered in their testament, as they charged their executors, as they pledged themselves, so do I possess. . . . Wherever diversity of doctrines is found, we will necessarily find adulteration of the scriptures and their interpretation. . . . What indeed has ever been found contrary to us in our Scriptures? What have we ever in-serted of our own for the sake of changing something in the Scriptures that we find inop-portune, whether by excerpting, supplementing or transforming [*detractione vel adiectione vel transmutatione*] them? . . . Marcion, on the other hand, openly and notoriously used not the pen but the expunging-knife, when he murdered the Scriptures in refashioning them to suit his purpose. . . . Though I read that "there must be heresies among us," these heresies could never have arisen without the Scriptures.

(Tertullian, *De praescriptione haereticorum*, chs. 38–39; ed. Refoulé, *Traité de la prescription*, 141–143). Similar arguments are found in Irenaeus and Jerome. Isidore of Seville borrows the passage for his *Etymologiae*. Tertullian alludes to 1 Corinthians 11:19, "For there must be also heresies among you, that they which are approved may be made manifest among you." On heresy and writing, see also Stock, *The Implications of Literacy*, 92–151.

56. Dionysius Thrax, *Technē grammatikē*, par. 5: ῥαψῳδία ἐστὶ μέρος ποιήματος ἐμπεριειληφός τινα ὑπόθεσιν. εἴρηται δὲ ῥαψῳδία οἱονεὶ ῥαβδῳδία τις οὖσα, ἀπὸ τοῦ δαφνίνῃ ῥάβδῳ περιερχομένους ᾄδειν τὰ Ὁμήρου ποιήματα. For ancient views on the rhapsodes and their activity, see Ford, "The Classical Definition of ῬΑΨΩΙΔΙΑ"; for an account of their role in establishing the Greek poetic canon, see Nagy, *Pindar's Homer*, 54–81, and *Poetry as Performance*, 61–79, 112–113. Although a minority opinion today, the derivation of "rhapsode" from "rhabdode" continues to be defended and curiously correlates with a preference for writing as the means of Homeric transmission: see for example Powell, *Homer and the Origins of the Greek Alphabet*, 216–217. For Milman Parry's view on rhapsodes, see "The Epic Technique of Oral Verse-Making, II," *The Making of Homeric Verse*, 337.

57. The scholiasts on Pindar seem to have acknowledged this problem by dividing the inheritance of Homer into more than one stream. "Originally, the name 'sons of Homer' was applied to Homer's descendants, who chanted his poetry in succession; later the rhapsodes no longer traced their lineage back to Homer, and those around Kynaithos became well-known: they inserted many epic passages [of their own] into the Homeric poetry. Now this Kynaithos was a Chian by lineage and was the first to rhapsodize the Homeric epics in Syracuse, around the 69th Olympiad [504–501 BCE],

according to Hippostratos." Drachmann, *Scholia vetera in Pindari carmina*, 3:28–32. On this passage and on the Homeridai generally, see Nagy, *Homer the Preclassic*, 59–73.

58. Stephanus, "In Poematia quae *Parōidai* sive *Parōidiai* vocantur Praefatio," *Homeri et Hesiodi certamen*, 71. "Consarcinatum" exactly translates συρραφεῖσα. On Henri Estienne, scholar and printer, see Renouard, *Annales de l'imprimerie des Estienne*.

59. Stephanus, *Certamen*, 71.

60. Ibid., 75.

61. For a reading of rhetorical theory as the working out of a subjacent concern with ownership, see Parker, "The Metaphorical Plot."

62. The bibliography on book culture is immense. For exemplary discussions, see Martin and Febvre, *L'apparition du livre*; Martin, *Histoire et pouvoirs de l'écrit*; Grafton, *Forgers and Critics*; and Johns, *The Nature of the Book*.

63. On skin and self, see Anzieu, *Le Moi-peau*. For an extension to medieval textuality, see Kay, "Original Skin."

64. *Καταχρηστικῶς* means "abusively" in ordinary discourse, but in rhetoric designates the trope of improper reference ("abuse" of language), and thus may here mean "at will, freely, without regard for the proper meaning." On the professional lineage of "Homeridai," see Deroy, "Le nom d'Homère."

65. On this quotation (*Aetia*, fragment 26) and its transmission see Pfeiffer, *Callimachus*, 1:35.

66. Ἀκέομαι, "to heal, mend, patch," is used of doctors, cobblers, and tailors, and often figuratively to describe the repairing of a wrong or an injustice. Here the reciters are depicted as "filling in" the gaps between the parts of the poem, as if the gaps betokened an unfortunate lack. See *LSJ*, s.v.

67. Drachmann, *Scholia vetera in Pindari carmina*, 3:28–31. The words in curly brackets { . . . } are added from another set of scholia (Mommsen, *Scholia recentiora*, 8) and may represent scribal additions. For another reading of these passages, see Nagy, *Poetry and Performance*, 112–113.

68. See *LSJ*, s.v.; citations related to drama and poetry are from Aristophanes and Lysias.

69. Much of Gregory Nagy's work is dedicated to recovering the event-character of ancient Greek poetry, particularly Homeric. See Nagy, *Pindar's Homer* and *Poetry as Performance*.

70. On the concept of "authority in performance" as the precursor of, and key to, that of "authorship in composition" in judging orally performed texts, see Nagy, *Poetry as Performance*, 19; on Homer as culture hero, see 74–79. For Nagy, "the transmitter as performer must also be authorized by his audience, who are presumed to be authoritative members of the song culture" (19). Claiming or challenging such authority might have happened in many ways, but not, in the period that interests us, by reference to a standard edition. Symmetrically, the standards the putative audience had in mind could be of many kinds, and may have changed over the generations.

71. Parry, "Homer and Homeric Style," 269. "At this point, let us quite forget the bookcases and libraries that nowadays preserve our studies, and be transported to other times and another world," commands Friedrich August Wolf as he begins to elaborate the hypothesis that epic was an orally composed genre (Grafton et al., trans., *Prolegomena to Homer*, 104; original text, *Prolegomena ad Homerum*, 71).

72. For analogous discussions, see Jakobson, "On Realism in Art" (1921), *Language in Literature*, 19–27; Bolter and Grusin, *Remediation*; Gitelman and Pingree, *New Media, 1740–1915*.

73. Jonathan Sterne, *The Audible Past*, 219; on questions of accuracy and credibility, see generally 216–225 and 236–261. See also Jütte, *A History of the Senses*. For an account of hearing as active exploration of an environment, see Gibson, *The Senses Considered as Perceptual Systems*.

74. On script reform, see Sterne, *The Audible Past*, 31–46. See Gitelman, *Scripts, Grooves, and Writing Machines*, for the tale of mid-nineteenth-century contests among longhand writing, various shorthand methods, and the early phonograph in the context of a mass print culture that could put to use increasingly exact transcriptions of legal proceedings, electoral speeches, parliamentary debates, etc.

75. Or to put it optimistically, see McLuhan, *The Gutenberg Galaxy*, 275: "And, as usual, when some previously opaque area becomes translucent, it is because we have moved into another phase from which we can contemplate the contours of the preceding situation with ease and clarity."

76. This formulation somewhat resembles Thomas Kuhn's notorious statements about the "incommensurability" of the objects envisioned under different scientific paradigms (see *The Structure of Scientific Revolutions*, 198–204). I should like to qualify the assertion as one of non-identity or non-substitutability rather than of incommensurability; in fact commensuration, successful or not, goes on all the time.

77. McLuhan, *The Gutenberg Galaxy*, 32. John Miles Foley, writing in 1998, updates the metaphor: literacy is to orality as a library is to the Internet ("The Impossibility of Canon," 18–20). Yet Gitelman observes that "Postmodernists such as Friedrich Kittler and Paul Virilio have been surprisingly romantic in their thinking about technology. I read a lingering determinism in their works" (*Scripts, Grooves, and Writing Machines*, 233).

78. *Odyssey* 1.325–360, 22.330–353, 8.72–108, 8.266–367, 8.471–541, 21.406–409; *Iliad* 8.185–189; *Odyssey* 4.269–290.

79. *Iliad* 7.175–189; compare *Odyssey* 19.386–394, 467–475. The Greek words translatable as "to read" are inventoried and richly commented upon in Svenbro, *Phrasikleia*. The prefix *ana-* ("up," "again," "back," "re-") is a striking common feature of many of these words, even when formed from distinct roots. On *anagignōskō*, see 9, 22, 30, 53, 61, 71, 182–184, 189, 200, 213, 217; on *anaphrazō*, see 21–23. For complementary discussions, see Nagy, *Greek Mythology and Poetics*, 202–222, and *Pindar's Homer*, 171–172.

80. *Iliad* 6.153–211. Excluding this episode (albeit with "regret") is a tradition begun by Rousseau; for a commentary, see Pop, *Antiquity, Theatre, and the Painting of Henry Fuseli*, 73–79. White ("Bellerophon in the 'Land of Nod'") argues for a Near Eastern origin, which would explain its inconcinnity.

81. For these *Iliad* and *Odyssey* examples, and an account of the latter poem's large-scale structure as determined by internalized media alternatives, see Saussy, "Writing in the *Odyssey*."

82. ἐχθρὸν δέ μοί ἐστιν | αὖτις ἀριζήλως εἰρημένα μυθολογεύειν (*Odyssey* 12: 452–453).

83. Létoublon, "Le messager fidèle," 123–124.

3. Autography

1. See Heubeck, "Schrift"; Powell, *Homer and the Origin of the Greek Alphabet*.

2. On the "ideographic" theme, see Derrida, *De la grammatologie*, 111–121, and for a discussion of the term, Saussy, *Great Walls of Discourse*, 45–57. On the cognitive steps preliminary to the origination of written signs, see Schmandt-Besserat, *Before Writing*. On children's acquisition of scriptive behaviors independently of letters, see Kress, *Before Writing*. On the varieties of writing systems, see Bright and Daniels, *The World's Writing Systems*.

3. Saussure, *Cours*, 150–151.

4. See Gomme, *An Historical Commentary on Thucydides*, 1:140–148.

5. See Gitelman, *Scripts, Grooves, and Writing Machines*, 21–61; I follow her history of shorthand and phonetic writing later. According to Ian Hunter, "Lengthy Verbatim Recall: The Role of Text," verbatim recording is imaginable only in a literate society. But see a striking example of indeterminate verbatim status on the nineteenth-century US frontier in Gitelman, *Always Already New*, 151. On the end-driven character of all transcribing, see Ochs, "Transcription as Theory."

6. Gitelman, *Always Already New*, 6.

7. Ellis, *A Plea for Phonotypy and Phonography*, 16–17. The terms "phonotypy" and "phonography" were coined by the shorthand expert Isaac Pitman.

8. Ellis, *The Alphabet of Nature*, 147.

9. Ibid., 148.

10. Ibid., 145; italics in original.

11. Bell, *Visible Speech*, 18–19.

12. Ibid., 20–21. Bell's "Visible Speech" is considered the direct ancestor of the International Phonetic Alphabet in use today. See John Kelly, "The 1847 Alphabet: An Episode of Phonotypy," 248–264, in Asher and Henderson, eds., *Toward a History of Phonetics*.

13. Cited in Bell, *Visible Speech*, 30.

14. Ibid., 22. One of those sons would have been Alexander Graham Bell, who in 1876 received the patent for the telephone.

15. Scott, "Principles of phonautography," tr. Feaster, in *Phonautographic Manuscripts*, 7; translation modified. The sealed packet was a cheap means of establishing priority in invention. Patent offices, unfortunately, paid more heed to those inventors with more energetic lawyers (Bell versus Gray, Edison versus Cros). On Scott generally, see Benoît et al., "Chronique d'une invention." Scott's original patent application and report to the Académie are visible at www.firstsounds.org.

16. Scott, "Note on phonautographic writing," trans. Feaster, in *Phonautographic Manuscripts*, 44.

17. Ibid., 51; Jean-François Ducis, *Othello*, in *Oeuvres* (Paris: Ledentu, 1839), 171. "In their cruel rage / Our desert lions, in their burning caves, / Are apt to rend shivering travelers limb from limb. / Better for him to let their devouring hunger / Rip apart the shreds of his quivering flesh / Than to fall alive into my terrible hands!"

18. Scott, *Phonautographic Manuscripts*, 57.

19. Nietzsche, "Über Wahrheit und Lüge im aussermoralischen Sinne, 1," *Sämtliche Werke*, 1:879–880 (referring to "Chladni figures"); Scott, *Phonautographic Manuscripts*, 24.

20. Rousselot, *Principes de phonétique expérimentale*, 136.

21. The earliest suggestion of a reversal of the causal direction of the phonauto-graph, making the traces move the stylus to produce vibrations in the membrane, oc-curs in Charles Cros's "Procédé d'enregistrement et de reproduction des phénomènes perçues par l'ouïe," 18 April 1877, in *Oeuvres complètes*, 579–580.

22. Benoît et al., "Chronique," par. 85.

23. On mechanically induced inscription as an "unreadable writing," a theory tested several times in the courts, see Szendy, *Écoute*, 97–108.

24. Writing or printing also involves an individual event—the contact of pen or printing plate with paper, for example. But our habits lead us to overlook this event except when exceptional circumstances bring it to the fore: calligraphy, signing and witnessing of legal documents, and the like.

25. On the hesitations of nineteenth-century law courts, sometimes recognizing in piano rolls and music boxes a form of "publication" (which therefore could be seen to infringe on copyright) and sometimes characterizing them as purely applied compo-nents of separate inventions, see Szendy, *Écoute*, 94–97, 102–104. Only when recordings were shown to be a means of reproducing words (in addition to sound), in a case of 1905, were they finally subordinated, by extension, to the law governing literary copy-right. Although an "unreadable writing," the tracks of the gramophone needle were now described as the "graphic notation of spoken words" (ibid., 103 n. 17): "graphic" in a sense including both Scott's invention and alphabetic writing.

26. See Rosen, "Researchers Play Tune Recorded Before Edison." Scott's singing can be heard at www.firstsounds.org/sounds/scott.php. For a fuller presentation of sonic "eduction," see Feaster, *Pictures of Sound*.

27. See Hoff and Geddes, "The Beginnings of Graphic Recording" and "Graphic Registration Before Ludwig"; also Brain and Wise, "Muscles and Engines."

28. On the contributions of the myograph and other apparatus to theory building, see Lenoir, "Helmholtz and the Materialities of Communication."

29. On this set of techniques generally, see Tufte, T*he Visual Display of Quantitative Information*. On their common artifactual basis, see Müller-Sievers, *The Cylinder*.

30. For the history of coordinate graphing, see Marey, *La méthode graphique*, 17–46, 109–111. For another graphic method meant to replace computation with visual recon-naissance, see Lalanne, "Description et usage de l'abaque."

31. Marey, *Du mouvement dans les fonctions de la vie*, 84–85.

32. Ibid., 95–96.

33. Marey, *Physiologie expérimentale*, vol. 1, iii.

34. Marey, *La méthode graphique*, i.

35. Ibid., vi.

36. Ibid., 51.

37. Ibid., *La méthode graphique*, 109; a point emphasized in Snyder, "Visibility and Visuality," 380–381.

38. Ibid., 107.

39. On the status of the body in Marey's and kindred writing systems, see Rous-seau, "Figures de déplacement." On "papers" as the output of science labs, see Latour and Woolgar, *Laboratory Life*. Marey's description of his own procedures anticipates Latour's "actor-network theory."

40. Marey, *La méthode graphique*, 341–342, discusses the phenakistoscope and the persistence of vision, so the elements of cinema were not far from his mind.

41. For the obstacles in the way of a smooth provincial career for Rousselot, see Baudet, "L'abbé Jean-Pierre Rousselot," 5–7.

42. Rousselot, *Les Modifications phonétiques*, 2.

43. Ibid., mentioning Tourtoulon et Bringuier, *Étude sur la limite*. Rousselot would have walked two or three hundred kilometers to perform this survey. St-Claud is clearly marked on the Tourtoulon map as belonging to the belt of "mixed" dialects that lies between the *langue d'oil* and *langue d'oc* zones. For the description of this "sous-dialecte des Marches," see *Étude sur la limite*, 42–51. The case of Tourtoulon, seen as a provincial intellectual crushed by the condescension and institutional power of Parisian mandarins, has been taken up again by Chevalier ("Le prophète et le roi"), invoking Bourdieu's model of "state nobility" and the monopoly of "the inheritors." But far from being a marginal man, Tourtoulon, despite his lack of a Parisian academic post, was a baron, a lawyer and a politician, and a proponent of a foreign policy based on alliances of "Latin" nations against the "Germanic ogre." Rousselot's linguistic advisor, Gaston Paris, is the main antagonist in Chevalier's story. For a history of dialect study in France, centering on Gilliéron and Edmont's ten-volume *Atlas linguistique de la France* (1902–10), see Desmet, Lauwers and Swiggers, "Le développement de la dialectologie française." On the importance of the Tourtoulon and Bringuier survey for later Prague School linguistics, see Sériot, *Structure et totalité*, 126–138.

44. Tourtoulon and Bringuier, *Étude sur la limite*, 6.

45. Gaston Paris, "Les parlers de France"; for the "wall" versus the "tapestry," see 163. Paul Meyer joined Paris in his distrust of clear "lines." On the mythic resonances of a unitary origin for French speech, see Bergounioux, "Le francien," and Cerquiglini, *Une langue orpheline*. Paris's lecture was printed in Rousselot and Gilliéron's dialectology journal (suspect, for Chevalier, of pro-centralizing bias). Jacobin linguistic centralism, most notably expressed in the Abbé Grégoire's "Report on the Necessity and the Means of Annihilating Patois and Universalizing the French Language" (1794), has been examined in de Certeau, Julia and Revel, *Une politique de la langue*. Tourtoulon and Bringuier, *Étude sur la limite*, 44–45, already urged the use of patois rather than the national idiom in primary school.

46. *Modifications* records an early differentiation into northern and southern dialects and cultural centers of gravity, and, on a smaller scale, particularities of dialect "coinciding for the most part so precisely with the parish bounds that [dialect differences] alone would nearly be sufficient to distinguish the different groups" (p. 348). But postmedieval language changes do not follow a North-South divide; they "go in a single geographic direction and seem to relate exclusively to the physical conditions of places and inhabitants" (ibid.).

47. Rousselot, *Modifications*, 2.

48. Rousselot, *Principes de phonétique expérimentale*, 1: 43.

49. On neogrammarians' approach to dialect, see Murray, "Language and Space."

50. Rousselot, *Modifications*, 170.

51. Ibid., 200–201.

52. Ibid., 351.

53. Ibid., 350–351.

54. Ibid., 351. In 1924, finding that the word "anemia" was too strong, he rephrased it to mean "a nervous weakening, imperceptible otherwise than through speech . . . which nonetheless I consider to be quite real" (*Principes*, 1108).

55. Rousselot, *Principes*, 45.

56. Rosapelly, "Inscription des mouvements phonétiques," 129. On this collabora-
tion, see Marey, *La méthode graphique*, 390–398; Brain, "Standards and Semiotics";
Teston, "L'oeuvre d'Etienne-Jules Marey"; Pisano, *Une archéologie du cinéma sonore*,
121–134.

57. Rousselot, *Principes*, 332.

58. Schwob, "La machine à parler" (1892), in *Oeuvres*, 246–249.

59. Rousselot, *Modifications*, 8.

60. Rousselot, *Principes*, 34–35.

61. The phoneticians' group at the Stockholm philological congress of 1886 had
affirmed that "languages should first be studied as they are spoken, not as they are
written. To make the spoken language accessible, it must be reproduced in writing with
conventional signs that represent sounds in a fixed and invariable manner." Quoted in
Passy, "Doctrines de la nouvelle école," in Bergounioux, *Aux origines de la linguistique
française*, 244.

62. Gauchat, "L'unité phonétique dans le patois d'une commune" (1905), 184–185.
For discussions, see Bergounioux, *Aux origines*, 144–145; Bonnet and Boë, "Émergence
de la phonétique générale et expérimentale," 51.

63. Bonnet et Boë, "Emergence," 52.

64. Ellis, *The Alphabet of Nature*, 148.

65. Gauchat, "L'unité phonétique," 231, cited in Desmet, Lauwers and Swiggers, "Le
développement de la dialectologie française," 36.

66. See Rousselot, *Principes*, 82, 100–102, 135–136.

67. Marey, *La méthode graphique*, vi.

68. Saussure, *Course*, 14; *Cours*, 30.

69. For "chimérique," see Saussure, *Cours*, 30. Significantly, it was at a moment of
theoretical despondency that Saussure told Meillet that only "the picturesque . . . the
ethnographic side of language" held any interest for him (de Mauro, "Notes biogra-
phies et critiques," in Saussure, *Cours*, 355).

70. Saussure, *Cours*, 164–166 (my translation; cf. *Course*, trans. Baskin, 118–120); for
other statements of the secondary status of material realizations, see *Cours*, 26, 98, 156
(*Course*, 10, 66, 112). On spatial and temporal difference and how they affect language,
see *Cours*, 261–289 (*Course*, 191–211). Saussure thought dialectology and linguistic
geography were important enough to be presented early in his classes, a fact obscured
by the chapter order imposed by the editors of the 1916 *Cours*. Framed by and centered
on the theory of the sign, the 1916 work thereby takes on a more internalist and synchronic cast.

71. Troubetzkoy, *Principes de phonologie*, 11–12.

72. Bloomfield, *Language*, 84, 127. One disciple of Rousselot seems to have made
a career out of flouting Bloomfield's advice. Giulio Panconcelli-Calzia, author of *Das
Als-Ob in der Phonetik* (1947), reportedly "never came to grips with the linguistic point
of view and he had no criteria for separating normal and deviant phenomena. His
theoretical outlook and his experimental and instrumental methods caused all phonetic categories (sound, syllable, etc.) to disintegrate. He turned them into fictions of
investigation without relevance to the communication process" (Kohler, "Three Trends
in Phonetics," 174–175).

73. Saussure, *Cours*, 156–157 (my translation); *Course*, 211.

74. Saussure describes the idealized alphabet in *Cours*, 165–166; compare the de-

scription, 64–65, of ancient Greek writing as ideally consistent and economical. Brain ("Standards and Semiotics," 280–286) sees Saussure as continuing the enterprise of laboratory phonetics and the graphic method; I see in the proclamation of the opposi-tive character of *langue* a sharp turn away from their results.

75. Saussure, *Cours*, 36–39. For an attempt to Saussurianize Rousselot, see Brain, "Standards and Semiotics," 280–284.

76. Pound, "Retrospect: Interlude," in *Polite Essays*, 129–130. On what Rousselot represented for Pound, see Golston, *Rhythm and Race in Modernist Poetry and Science*, 64–73, 113, 136. On Pound and media, see Tiffany, *Radio Corpse*; Campbell, *Wireless Imagination*.

77. Mallarmé, "La musique et les lettres," *Oeuvres complètes*, 643.

78. See Auroux, *Histoire des idées linguistiques*, 3:503–504.

79. On Hugo's versification, see Tamine, "Le vers de *La Légende des Siècles*."

80. On the 1880s context of prosodic invention, see Boschian-Campaner, *Le vers libre dans tous ses états*; Bobillot, "René Ghil."

81. From a large bibliography, see Dujardin, *Les premiers poètes du vers libre*; Kahn, *Premiers poèmes*; Taupin, *L'influence du Symbolisme français*, 112–124; Bergeron, "A Bugle, a Bell, a Stroke of the Tongue"; Sieburth, "The Work of Voice"; Brain, "Geneal-ogy of 'Zang Tumb Tumb.'"

82. Cited, de Souza, *Du rythme en français*, 10.

83. de Souza, *Du rythme en français*, 8, 64. De Souza is thinking of Rousselot's axiom that we never hear ourselves actually speaking, only as we imagine ourselves to be speaking.

84. Ibid., 14–15.

85. Ibid., 15–18. For Rousselot's version, see "La phonétique expérimentale: Cours professé au Collège de France par M. l'abbé Rousselot, leçon d'ouverture," 21.

86. Taupin, *L'influence du Symbolisme*, 116. Compare the opening of poetry to other registers of sound with the coming of the tape recorder, as narrated by Scott, "Re-Conceiving Verse."

87. *Dictionnaire de l'Académie française*, 7th edition (1877), s.v. *Alexandrin, -ine*.

88. Lote, *L'Alexandrin*, 16–17, 5–6. The poet André Spire built on Lote's research in *Plaisir poétique et plaisir musculaire* (1943, 1986).

89. Ibid., 16.

90. Ibid., 453.

91. Ibid., 460–461.

92. Ibid., 466.

93. Ibid., 539.

94. Ibid., citing Rémy de Gourmont, *Esthétique de la langue française*.

95. Ibid., 566.

96. Ibid., 701.

97. For a history of experimental studies of English verse and accentuation in the Rousselot tradition, see Hall, "Mechanized Metrics."

98. Bogatyrev and Jakobson, "Folklore as a Special Form of Creativity," 32; Trou-betzkoy, "La phonologie actuelle," 233.

99. "Situated on the border between organic human and inorganic machine, [the phonoscope] registers the mutually informing rhythms of the biological and mechani-cal worlds." Golston, *Rhythm and Race*, 72.

100. Helmholtz, *On the Sensations of Tone*, 103–104; see also 43–45, 119–129, and the

translator's additions (538–542) bringing Helmholtz's 1862 book up to date as of 1885. For a current view on the physics of vowel production, see Hardcastle and Laver, *The Handbook of Phonetic Sciences*, 72–73.

101. On Koenig's life and career, see Pantalony, *Altered Sensations*. After Koenig's death, Rousselot acquired for the Collège de France Koenig's "Grand Tonomètre" consisting of 672 calibrated tuning forks ranging from 32 to 45,000 Hertz, well below and above the human hearing range (description in Koenig, *Catalogue des apparels d'acoustique*, 19–20).

102. Pernot, "L'abbé Rousselot," 19.

103. For the figure of 912 Hz, see Rousselot, *Principes*, 746.

104. Ibid., 1105.

105. The "wave theory" of linguistic change, proposed by Johannes Schmidt in 1872, is discussed in Saussure, *Cours*, 282–289 (*Course*, 209–210).

106. Gilliéron, *Les étymologies des étymologistes et celles du peuple*, 65. Gilliéron and Rousselot had little in common besides an interest in dialect, yet they use the same metaphors of vibration and milieu. On this passage, see Desmet, Simoni-Aurembou and Swiggers, "Introduction," in Lauwers, Simoni-Aurembou and Swiggers, eds., *Geographie linguistique et biologie du langage*, 10.

107. Rousselot, *Modifications*, 351–352.

108. Ibid., 350 (italics in original).

109. Bergson, *Matière et mémoire*, in *Oeuvres*, 255 (my translation). "Schème moteur" is lamentably rendered in the reigning English translation of this work as "motor diagram." The problem of segmentation arises in Saussure, *Cours*, 145 (*Course*, 104), where it makes the transition to the theory of the bimaterial sign (composed of phonetic substance and semantic substance in an arbitrary relation). The metaphor of a continuous ribbon—and the word "synchronic"—may derive from memories of phonautograph tracings. (See *Course*, ed. Meisel and Saussy, 237.)

110. Bergson, *Matière et mémoire*, 255.

111. Saussure, *Cours*, 162 (*Course*, 117).

112. Stetson, *Motor Phonetics*, 5.

113. Ibid., citing Rousselot, *Principes*, 1:335. Stetson grew increasingly bitter over time at the "confusion" induced, as he put it, "when the attempt is made to treat an articulate language like a paper language . . . when the phonemicists posit nothing but the differentiation of symbols and their linear arrangement in order to construct a phoneme system in terms of 'oppositions' and 'positions.' . . . This 'method of opposition' was to bulk large and to be extended by Trubetzkoy to all linguistics," betraying, in Stetson's view, the legacy of Saussure, "too keen and practical a phonetician to identify an individual unit in a system by mere unlikeness" (*Bases of Phonology*, 101, 22–23, 9, 27). Stetson clung to Rousselot's kind of linguistics well into the 1950s. For recent work drawing on Stetson's model, see Liberman and Mattingly, "The Motor Theory of Speech Perception Revised." On the Stetson-Rousselot tradition as a factor in modern musical composition, see Maconie, "The French Connection."

114. Bergson, *Matière et mémoire*, 223–224.

115. Ibid., 180–181.

116. Jay, *Downcast Eyes*.

117. See, e.g., *Jenseits von Gute und Böse* (*Sämtliche Werke*, 5: 27, 29, 38), *Der Fall Wagner* (6:418).

118. Ribot, "Les mouvements et leur importance psychologique," 373. Compare

Dewey, "The Reflex Arc Concept in Psychology." On the revisionary force of the idea of nervous reflex, see Gauchet, *L'inconscient cérébral*, 41–68.

119. Ribot, "Les mouvements et leur importance psychologique," 384; italics in original. Henry Maudsley may be the originator of the theory that memory encodes movement: see *The Physiology of Mind*, 463–470.

120. James, "What Is an Emotion?" 190.

121. Richet, "De l'influence des mouvements sur les idées," 614. On backflow, see Kittler, *Gramophone, Film, Typewriter*, 38.

122. Ribot, *La vie inconsciente et les mouvements*, 1–2, ii–iii, 23, 32–33, 77.

123. See Laura Otis, "The Metaphoric Circuit"; Brain and Wise, "Muscles and Engines."

124. For examples of early physiological rhythm research, see Bolton, "Rhythm"; Bücher, *Arbeit und Rhythmus*; Fraisse, "Contribution à l'étude du rythme en tant que forme temporelle"; Hall and Jastrow, "Studies of Rhythm"; Landry, "Le Rythme musical"; MacDougall, "The Structure of Simple Rhythm Forms"; Stetson, "Rhythm and Rhyme," "A Motor Theory of Rhythm and Discrete Succession"; Swindle, "On the Inheritance of Rhythm." Abraham, *Rythmes de l'oeuvre, de la traduction et de la psychanalyse* begins with an evocation of these quantitative studies. For a general historical discussion, see Michon, "Notes éparses sur le rythme."

125. Woodworth, "Non-Sensory Components of Sense Perception," 171.

126. Wallaschek, "On the Difference of Time and Rhythm in Music," 33.

127. Guyau, "La mémoire et le phonographe," 319–320; translation from Kittler, *Gramophone, Film, Typewriter*, 30.

128. Ibid., 320, 322; 30, 32 (translation modified).

129. Ibid., 322; 32 (translation modified).

130. Kittler, *Gramophone, Film, Typewriter*, 33.

131. Ribot, *Les maladies de la mémoire*, 94–95. On this work, see Nicolas, *La mémoire et ses maladies selon Théodule Ribot*.

132. Ribot, *Les maladies de la volonté*, 151–152.

133. Janet, *L'automatisme psychologique*, 48.

134. Ibid., 52. Janet's study involved fourteen hysterical women, five hysterical men, and eight other patients with epilepsy or other nervous disorders. Since his star patients were all women, I use feminine pronouns in translating the nongendered pronouns of the original.

135. Ibid., 53–54.

136. See Roussel, *Locus Solus* (1914), 146–226, and for details of Roussel's case, Janet, *De l'angoisse à l'extase*, 2:132–136; Kahn, "Death in Light of the Phonograph"; Garrabé, "Martial, ou Pierre Janet et Raymond Roussel."

137. Janet, *L'automatisme psychologique*, 69.

138. Ibid., 72–73.

4. The Human Gramophone

1. André, "Psychologie expérimentale et exégèse," *La Croix*, 3 February 1927, 4. The sentence "We have the words of Jesus" recurs often in Jousse (e.g., in a lecture of April 6, 1937, in *Dernières dictées*, 232).

2. The thesis that an oral text has no "original" dawned on Parry only in the course of his fieldwork in Yugoslavia. See also Aloys de Marignac, "Esquisse d'une nouvelle

méthode de critique homérique," which cites Jousse but not Parry in characterizing the search for an oral Ur-text as fruitless.

3. André, "Psychologie expérimentale et exégèse," 4.

4. On the crisis, see Poulat, *Histoire, critique et dogme dans la crise moderniste*; Hill, *The Politics of Modernism*; Laplanche, Biagioli, and Langlois, *Autour d'un petit livre*; Goichot, *Alfred Loisy et ses amis*.

5. "Truly the historian's task of distinguishing what is traditional and what is peculiar, what is kernel and what is husk, in Jesus' message of the kingdom of God is a difficult one and laden with responsibility" (Harnack, *What Is Christianity?* 60 [translation modified]; *Wesen des Christentums*, 36). One hundred years earlier, and for similar reasons, Schleiermacher (*Über die Religion* [*On Religion: Speeches to its Cultured Despisers*], 1799) had made a similar attempt to distinguish the wheat and the chaff.

Harnack was an eminent historian of dogma, named by Bismarck to a chair at Berlin over the objections of the Lutheran hierarchy. The eleventh edition of the *Encyclopaedia Britannica* (1911) catches him at his apex: "His distinctive characteristics are his claim for absolute freedom in the study of church history and the New Testament; his distrust of speculative theology, whether orthodox or liberal; his interest in practical Christianity as a religious life and not a system of theology" (s.v. "Harnack, Adolf"; see also Liebing, "Adolf von Harnack"). Ennobled in 1914, he was the first director of the Kaiser-Wilhelm Institut, Germany's leading research foundation.

6. Harnack, *What Is Christianity?* 55; *Wesen des Christentums*, 33.

7. Ibid., 69; 41.

8. "Kant" is used here as shorthand for the demand that religious claims be justifiable "within the limits of reason alone." Thomas Jefferson's selection from the Gospels and Hegel's youthful retellings of the life of Jesus are other examples of this demand, although Jefferson is unlikely to have read Kant (see Jefferson, *Extracts from the Gospels*, and Hegel, *Theologische Jugendschriften*).

9. Harnack, *What Is Christianity?* 226–228, 253; *Wesen des Christentums*, 132–133, 145.

10. Ibid., 289 (referring to Luther); 168.

11. Loisy, *L'Évangile et l'Église*, xx.

12. Ibid., xx–xxi.

13. Loisy, *Autour d'un petit livre*, vii–viii.

14. Loisy, "La question biblique et l'inspiration des Écritures," cited in Goichot, *Alfred Loisy et ses amis*, 28.

15. Poulat, *Histoire, dogme et critique dans la crise moderniste*, 44. Loisy's "project [at the time of his first censure in 1893] was not, like Renan's, tied to an absolutization of reason, but rather to a Catholic counteroffensive on the terrain of historical biblical criticism. Polemics and persecutions radicalized Loisy by putting him before the ultimate choice of criticism or Catholicism" (Jérôme Grondeux, "Georges Goyau et le modernisme," in Laplanche, Biagioli, and Langlois, *Alfred Loisy cent ans après*, 140).

16. "Holy Office," i.e., the Inquisition, known today as the Congregation for the Doctrine of the Faith. On the process of review and condemnation, with reproduction of original documents, see Arnold and Losito, *La censure d'Alfred Loisy*. Loisy's chief antagonists were Cardinal Richard, archbishop of Paris; Louis Billot, S.J., "the well-known agent of a severe and intransigeant Thomism" (*La censure*, 37); Pie de Langogne, Consultor of the Index and the Holy Office; and Rafael Merry del Val, papal secretary of state.

17. "Lamentabili sane exitu" (Pius X, 3 July 1907), in Denziger, *Enchiridion*, 669–674.

18. "Pascendi dominici gregis" (Pius X, 8 September 1907), in Denziger, *Enchiridion*, 675–683; letter *motu proprio* "Sacrorum antistes" (Pius X, 1 September 1910), ibid., 688–690. On the application of these measures, see Laplanche, *La crise de l'origine*, esp. 41–70.

19. Colin, *L'audace et le soupçon*, 237; see also Goichot, *Alfred Loisy et ses amis*, 90–91. As de Vitry says of the 1864 "Syllabus errorum" appended to the encyclical "Quanta cura" (Pius IX, 8 December 1864; Denziger, *Enchiridion*, 574–584), it gives the endangered Church an identity as the negation of her own negation ("Catholicisme et représentation," 5).

20. "The modern spirit, that is, rationalism, criticism, liberalism, was founded the same day philology was founded. *The founders of the modern spirit are the philologists.*" Renan, *The Future of Science*, 133 (*L'avenir de la science*, 141; italics in original). On the stigmatization of Judaism as rationalism and rationalism as Judaizing, see Nirenberg, *Anti-Judaism*, 372–380.

21. See Houtin, *La question biblique chez les catholiques de France au XIXe siècle*. The encyclical "Qui pluribus" (Pius IX, 9 November 1846, excerpted in Denziger, *Enchiridion*, 556) had denounced "the error of rationalism"; "Providentissimus Deus" (Leo XIII, 18 November 1893; Denziger, 638–642) opposed the "higher criticism" with the assertion that the Scriptures are unerring and inspired. The Pontifical Biblical Commission, founded in 1903 and subordinated to the Inquisition, was designed to resolve all scholarly questions relating to the Scriptures. Its first head was the relatively accommodating Cardinal Rampolla, freshly displaced as secretary of state by the more activist Merry del Val (Arnold and Losito, eds., *La censure d'Alfred Loisy*, 49). Not until "Divino afflante Spiritu" (Pius XII, 30 Sept 1943; Denziger, *Enchiridion*, 754–757) was a greater latitude permitted in biblical studies.

Loisy's applications of criticism are not particularly bold, except in a Catholic context. He argues, for example, that Mark must have been the main source used by the composers of Matthew and Luke, who also benefited from a "sayings source" not otherwise transmitted; he emphasizes the narrative and doctrinal distinctiveness of John; he sees the predictions of the destruction of the Temple (which occurred in 70 CE) as having been inserted post factum (*L'Évangile et l'Église*, 3–15). None of these points is original with Loisy; he is expressing a consensus well established by 1902. See Kümmel, *Introduction to the New Testament*; Farmer, *The Synoptic Problem*.

22. Loisy, *L'Évangile et l'Église*, xviii–xix; see also *Simples réflexions*, 55. These very matters had been disputed in France by Simon and Bossuet more than two hundred years before; see *Autour d'un petit livre*, 23–39.

23. Loisy, *Simples réflexions*, 132.

24. Abbé Gaston Oger, *Évangile et évolution: Simples remarques sur le livre de M. Loisy* (Paris: Téqui, 1903), cited in Poulat, *Histoire, critique et dogme dans la crise moderniste*, 147. Emile Berliner registered the "His Master's Voice" trademark and slogan in 1900. Loisy calls on the same figural register in arguing that the life and contingency of history bleed away if the only legitimate interpretation of texts is to see them as confirming the truths of doctrine: "as if Christ, that divine automaton, had done nothing but to perform at every moment the words and actions specified by a program predetermined in all its details, where the future was no less exactly defined than the present; as if the apostles had been swept up in the same determinism, and the Church came into the world by a mere play of supernatural mechanics" (*Autour d'un petit livre*, 18–19).

25. Quoted, Baron, *Mémoire vivante*, 67. The editor of the series "Archives de Phi-losophie," Pedro Descoqs, S.J., was a prominent supporter of Maurras and the Action Française until its condemnation by the Vatican in 1926. On his opposition to Blondel's philosophy, see Bernardi, *Maurice Blondel, Social Catholicism, and Action Française*.

26. André, "Psychologie expérimentale et exégèse," *La Croix*, 3 February 1927, 4.

27. On Billot's contribution to the case against Loisy, see Arnold and Losito, *La censure*, 37–39, 153–170. On his particularly uncompromising line on the divine inspira-tion of the Bible, see Laplanche, *La crise de l'origine*, 67–69. Billot resigned as cardinal, an unusual step, after the condemnation of the Action Française.

28. Remarks cited in Baron, *Mémoire vivante*, 74–76. On Frey, see Laplanche, *La crise de l'origine*, 286–287; on Fonck, 53–54.

29. The supposition is not in itself exorbitant. Black finds that in the Gospels, the passages that most clearly hint back to an Aramaic original are the words of Jesus; this suggests that the Aramaic basis would have been a logia text, and the narrative parts supplied by authors more familiar with Greek (though their Greek often shows signs of translationese): Black, *An Aramaic Approach to the Gospels and Acts*, 271–277.

30. Luke 1:39–56 (KJV); italics mine.

31. Loisy, "Sur l'origine du Magnificat," 428; the reading with no proper name is admitted as a conjecture in the apparatus of the Nestlé-Aland *Novum Testamentum graece*.

32. I Samuel 2:1–9 (KJV).

33. Loisy, "Sur l'origine du Magnificat," 431–432.

34. I Samuel 2:1, in the Vulgate version.

35. Psalm 103 (Vulgate numbering), verse 1.

36. Psalm 12, verse 6.

37. Psalm 145, verse 2.

38. Psalm 26, verse 9.

39. Psalm 33, verse 4.

40. Psalm 143, verse 7.

41. Luke 1:46–47. Jousse, *Le style oral*, 110.

42. See Jousse, *L'anthropologie du geste*, 293–295; *La manducation de la parole*, 71–74.

43. Jousse, *Le style oral*, 143 (italics mine).

44. See the reviews of Baron, *Mémoire vivante*, by Guillaumont and Poulat. The most significant contemporary reviews of *Le style oral* were signed by Meillet, Grand-maison, and Loisy.

45. Bremond, *La poésie pure*, 11, 148; *Prière et poésie*, xiv–xv. On the debate, see Valéry, "Discours sur Henri Bremond," *Oeuvres* 1:763–769, and Arnold, "La querelle de la poésie pure."

46. Lefèvre, *Marcel Jousse: Une nouvelle psychologie du langage*; "Une nouvelle psychologie du langage." In *L'adhésion* (1943), a meditation on the illocutionary force of language, Lefèvre is still considering Jousse's mnemonic rhetoric (349–354).

47. The journal *Les Lettres* was particularly aggressive in advancing this claim. A version of Lefèvre's essay, with the claim about the words of Jesus set in boldface, was inserted like a wedge between parts one and two of a jaundiced account of Loisy's jubilee celebration at the Collège de France, entitled "A Jubilee Without Jubilation" and adorned with caricatures. See *Les Lettres* 14:2 (June 1927): 129–140, 141–170, esp. 169, and 14:3 (July 1927): 258–293.

48. Lefèvre, *L'itinéraire philosophique de Maurice Blondel*, 89–102, commented in Bremond and Blondel, *Correspondance*, 3:315–316.

49. Jacquin, *Notions sur le langage*.

50. See André Ombredane, "Le langage," 3:363–458 in Dumas, *Nouveau traité de psychologie*, esp. 374–375.

51. Grandmaison, *Jésus-Christ*, 1:202. The appreciation of Jousse continues to 209. A second posthumous edition omitted the praise of Jousse.

52. Loisy, *La naissance du christianisme*, 49. The prior article is "Le style rythmé du Nouveau Testament," 1923. On Jousse's apparent "outflanking" of Loisy, see Jones, *Independence and Exegesis*, 102–105.

53. Jousse, "Les récitatifs rythmiques"; on James Joyce's attendance at one of these performances, see Colum and Colum, *Our Friend James Joyce*, 130–131; Heath, "Ambiviolences," 56; Burns, *Gestural Politics*, 161–165.

54. For the stenographic transcripts of these three series of lectures, now in digital form, see Jousse, *Cours oraux*.

55. "Rhythmocatechist," *Time*, 6 November 1939, 54.

56. See Torgovnick, *Gone Primitive*.

57. On the sudden conversions of Ernest Psichari, Raïssa Maritain, Jacques Maritain, Jacques Rivière, Jean Cocteau, and others, see Schloesser, *Jazz Age Catholicism*.

58. Maritain, *Antimoderne*, 15; and see Compagnon, *Les antimodernes*, 245–252.

59. Schloesser, *Jazz Age Catholicism*, 15–16; on the year 1926, see 131–135.

60. Lefèvre, "Une nouvelle psychologie du langage," 70.

61. Garron, "La part typographique," 62, on Mallarmé's "Coup de dés."

62. Foley, *The Singer of Tales in Performance*, 67–68, on Dennis Tedlock's transcription of Zuñi stories. Where I have inserted "imagined," Foley wrote "heard."

63. Jousse, "Les formules targoumiques du 'Pater,'" 122, on Matthew 13:52 (KJV): "Therefore every scribe *which is* instructed unto the kingdom of heaven is like unto a man *that is* an householder, which bringeth forth out of his treasure *things* new and old."

64. On this concept, see Sebban, "La genèse de la 'morale judéo-chrétienne.'"

65. For a more recent attempt to make the bridge between oral interpretation of Torah and the "ministry of the word," see Gerhardsson, *Memory and Manuscript*.

66. Jousse, *L'anthropologie du geste*, 292 (midrash on Matthew 11:28–30).

67. Ibid., 301.

68. Jousse, *L'anthropologie du geste*, 313.

69. Ibid., 216, on Matthew 18:3.

70. Jousse, *Le parlant, la parole et le souffle*, 166.

71. Ibid., 273–274, midrash on John 15:10, 13.

72. "Iōhānān bar Zabdaï of Galilee" (i.e., Saint John), cited in ibid., 270. In *La manducation de la parole*, this citation is used to confute those who detach John from the Synoptics—an impossible separation, for Jousse, because "this preferred learner, the one taught by predilection, will be precisely the one who will learn and preserve the *superior rhythmo-catechism*, which some suppose him to have stylistically and rhythmically remodeled" (99). Here two arguments assuming the inerrancy of scripture ("the Gospel of John is authentic" and "the author of this Gospel was 'the disciple whom Jesus loved'") combine with an argument about the errors of translation to yield precisely the history Jousse needs.

73. Jousse, *Le parlant, la parole et le souffle*, 205 (internal reference to "Targ. Job, 16,20; 33, 23").

74. Ibid., 113.

75. Ibid., 115.

76. Jousse, *L'anthropologie du geste*, 299–300, on Matthew 22:37.

77. Ibid., 312.

78. Ibid., 322, on Matthew 28:19.

79. On the École d'Anthropologie—a private foundation established in 1876 and always in discreet concurrence with the public universities and the museums of natural history—see Conklin, *In the Museum of Man*, 28, 33, 44, 52–60. Jousse's professorship there was endowed in 1932 by Emmanuel Desgrées du Loû, a Breton lawyer and cofounder of the regional newspaper *L'Ouest-Éclair* (later *Ouest-France*). The paper's politics were democratic and social-Catholic; it opened at the height of the Dreyfus Affair and took the captain's side. On Desgrées du Loû, see de Cadore, "*L'Ouest-Éclair* et les deux Ralliements" and Delbreil, "*L'Ouest-Éclair* et le Parti Démocrate Populaire." Louis Marin, president of the École from the 1920s to the 1950s, was a prominent conservative politician and a close associate of Desgrées du Loû through the PDP. On Marin, see Lebovics, *True France*, 12–50.

80. Jousse, *Cours oraux*, École d'Anthropologie, 7 November 1932.

81. Montandon, "Le squelette du Professeur Papillault." On the autopsy movement among materialists in Third Republic France, see Hecht, *The End of the Soul*. Montandon, a firm eugenicist, persuaded the occupying Germans to establish him as the editor of a new journal about the French "ethnie" (≈ *Volk*), and performed anthropometric inspections, delivering certificates of Jewish or non-Jewish status. The Resistance killed him in August 1944 (Conklin, *In the Museum of Man*, 170–186, 308–325).

During the Occupation, the École's activities were spotty. Its president, Louis Marin, had left the country to join de Gaulle. On his return he declared, in de Gaullian fashion, that the École had simply shut down in order to avoid contact with the occupant. But lectures had continued in wartime, temporarily housed in the Faculty of Medicine, and volume 51 of the house journal, the *Revue anthropologique*, contained "L'anthropologie du geste et les proverbes de la Terre" (an unoriginal summary of Jousse's 1941 lectures by a student named Adolphe V. Thomas), immediately followed by Henri Briand's glowing account of a speech on "German Racial Legislation" by Professor Eugen Fischer of Berlin, "the Third Reich's principal technician for Eugenics and Heredity." Presumably Marin would not recognize these activities as being in the continuity of the École.

82. Conklin, *In the Museum of Man*, 44.

83. Papillault, *Des instincts à la personnalité morale*, 146. Papillault also developed his highly biologized version of Freudian theory, "biopsychanalysis."

84. Cited by Conklin, *In the Museum of Man*, 54.

85. Jousse, *Le style oral*, 155, 234 (italics mine).

86. Jousse, *L'anthropologie du geste*, 206.

87. See Ghesquier-Pourcin, ed., *Energie, science et philosophie*. Aby Warburg's program of art-historical research took from Wundt, Fechner, Schmarsow, and Semon a model of memory as stored energy. See Roland Recht, "Introduction," in Warburg, *L'atlas Mnémosyne*, 19, 22–23, 40–21.

88. For examples, see Stéphanie Dupouy, "Charles Féré (1852–1907) et l'énergétique du système nerveux," 291–305 in Ghesquier-Pourcin, *Energie, science et philosophie*; Ri-

bot, *La vie inconsciente et les mouvements*, 77, 149; d'Udine, *L'art et le geste*, viii–ix. This commitment to mimesis distinguishes Jousse from structuralism, despite similarities (de Certeau, "Une anthropologie du geste: Marcel Jousse").

89. d'Udine, *L'art et le geste*, xiii–xvi. The work is dedicated to the music educator Émile Jaques-Dalcroze and the biologist Félix Le Dantec (the former having contributed the idea of rhythmic gesture, the latter that of rhythmic vibration of colloidal tissue). On Haeckel's development of similar ideas, see Brain, "Materialising the Medium," 128.

90. Maurras, *Le chemin de Paradis*, preface. Subsequent editions removed the most offensive sentences; they may still be viewed at http://maurras.net/2012/07/12/preface-du-venin/. Jousse objects to this passage frequently in his lectures: e.g., École d'Anthropologie, 11 March 1941, 24 February 1942. Other churchmen such as Pedro Descoqs, S.J., understood Maurras as rejecting only the "dangerous interpretations" of the Gospel in the hands of "revolutionaries and democrats" (see Bernardi, *Social Catholicism*, 104).

91. Frey, "Le Pater est-il juif ou chrétien?" 563.

92. Frey, "Le Pater," 556. Harnack's investigation of the Lord's Prayer, "Die ursprüngliche Gestalt des Vater-Unsers," "Über einige Worte Jesu," 195–208, does not mention "odor." On the rich tradition of *foetor judaicus*, see Geller, *The Other Jewish Question*, 273–297.

93. Frey, "Le Pater," 557.

94. Ibid., 561.

95. Ibid., 562; italics in original. For the "moral elevation," Frey cites, with undisguised irony, Loisy. Harnack saw in Luke's version hints of an initiation prayer "through which the status of Christian is first founded" ("Die ursprüngliche Gestalt," 205).

96. Baron, *Mémoire vivante*, 100 (allegedly said to Frey in Rome during Jousse's 1927 visit). For a more recent view, see Vermes, *The Religion of Jesus the Jew*, 162–167.

97. Vaugeois, "L'Action française," 17, 18, 23, 25.

98. Ibid., 25.

99. Serry, *La naissance de l'intellectuel catholique*, traces the opportunities for influence opened to the Action Française by the ban on Modernism in 1907 and the condemnation of "Le Sillon" in 1910 (62, 152). See Sutton, *Nationalism, Positivism and Catholicism*, 145–153, 182–191, 257, for an account of the rivalries among these tendencies.

100. The reason had nothing to do with the anti-Semitism of Action Française and a great deal to do with the merely instrumental role it assigned to Roman religion in creating order: "Politics first!" A first condemnation drafted in 1915 was kept secret by Pope Pius X. On the overt condemnation of 1926, see Weber, *Action Française*, 230–239; Bernardi, *Maurice Blondel*, 208–217; Serry, *Naissance de l'intellectuel catholique*, 267–292.

101. See Connelly, *From Enemy to Brother*, 7, 97, 182. The prayer was removed only in 1959. On its meanings, see Oesterreicher, "Pro perfidis Judaeis." On relations between the Catholic Church and Jewish institutions in France, see Dujardin, "Les relations entre chrétiens et juifs depuis 50 ans" and Pierrard, *Juifs et Catholiques français*.

102. The remark is reported in Bonnefoy, *Vers l'unité de croyance*, 20. The editor of *Études* wrote a long narrative review of Pallière's autobiography: see Grandmaison, "L'odyssée spirituelle d'un Moderniste." On another priest suspected of Modernizing

and Judaizing for his study of the Targums, see Deffayet, "Le Père Joseph Bonsirven S.J."

103. On the Semaine, held annually from 1921 to 1928, see Serry, *Naissance de l'intellectuel catholique*. The 1927 event was strongly marked by the condemnation of the Action Française the previous year. Its conciliatory theme betokens the weakened position of members of the board previously associated with the AF and integralism (Serry, 283). But one effect of the condemnation was to demoralize much of the leadership of the "Catholic cultural renaissance": the festival and the journal that sponsored it, *Les Lettres*, soon collapsed.

104. Halflants, "Le Sémitisme chez les 'Écrivains Catholiques.'"

105. Ibid. For two other accounts of the evening, see Bernoville, "Après la 'Semaine'"; Pierrard, *Juifs et Catholiques français*, 251–253.

106. Colmet, "L'abolition des 'Amis d'Israël,'" 384. An integralist Catholic website, now no longer in existence, in 2012 was still linking Jousse with "the accursed Jewish conspiracy": see "M.V.," "Père Jousse et la maudite conspiration juive."

107. They appeared in the *Revue juive de Genève* and *Cahiers juifs*, an international quarterly for exiles and those considering exile. On *Cahiers juifs*, see Saussy, "The Refugee Speaks of Parvenus."

Jousse acknowledges the attention he receives from Jewish periodicals by citing Meillet's phrase: "possibly one of the greatest intellectual joys of my life." *Cours oraux*, Hautes études, 15 May 1934.

In the usual works on Jousse, Boucly is represented as "Étienne Boucly" or "Bougly," which suggests that she was not known to the keepers of the flame. A file for her award of the Legion of Honor (number 259.013, dated 11 April 1952), with slender biographical details, subsists in the Archives Nationales.

108. Boucly, "Israël, auteur classique," 39.

109. Boucly, "La mimique hébraïque," 204–205, 207. After moving to Palestine, Bialik became the first major poet in modern Hebrew. The pages of *Cahiers juifs* frequently present his poems and plays in French translation.

110. Meschonnic, *Critique du rythme*, 689, 698.

111. See Jousse, *Cours oraux*, Hautes études, 7 March 1939: "I say intellectual encounter. Let me seize the moment to explain: I've been told that I refused to convert Jews and that that vexed them enormously. I'd rather say that I'd sooner vex the Jews than try to convert them. There are plenty of religious organizations out to convert Jews, especially now, when, as it appears, there are so many Jews seeking to convert that the Catholic hierarchy has had to issue regulations for their conversion. The Jews are in such a state of suffering and exile that I would not very much like to see them convert to Catholicism at such a moment. My position: I consider that my role is to present my science objectively, without the Jews among you thinking that I'm laying an ambush." See also Hautes études, 30 April 1935, 1 June 1937, 28 November 1938, 21 March 1939; École d'Anthropologie, 28 February 1949. One of the priests most active in the effort to convert Jews, Joseph Bonsirven, S.J., was a longtime antagonist of Jousse within the Society. On his career, see Deffayet[-Loupiac], "Le rôle du Père Joseph Bonsirven" and "Le Père Joseph Bonsirven."

112. Jousse, *Cours oraux*, École d'Anthropologie, 11 March 1941, 1–9. The "French organization" was the Action Française.

113. On the development of these ideas, see Lebovics, *True France*. "In reply to all that [i.e., bookish civilization], Pétain has said: you are going to become a peasant once

more. It's hard to go back. The return to the earth means: the return to normal hands
. . . the return to normal objects . . . the return to the normal sense of words" (*Cours oraux*, École d'anthropologie, 5 January 1941, 16).

114. Jousse, *Cours oraux*, École d'Anthropologie, 11 March 1941, 12–15.

115. Ibid., 5 January 1941, 7.

116. Ibid., 24 February 1942, 2–3.

117. See Faure, *Le projet culturel de Vichy*; Pierrard, *Juifs et catholiques français*, 299–305; Conklin, *In the Museum of Man*, 317–324.

118. See, e.g., McLuhan, *The Gutenberg Galaxy*, 18–21, 33–35, 45–47.

5. Embodiment and Inscription

1. Plato, *Phaedrus*, 277e–278 a. For examples of the traditional reading of the antithesis, see Ficino, *Commentaries on Plato: Phaedrus and Ion*, 190–192; Griswold, *Self-Knowledge in Plato's Phaedrus*, 204–214. A similar antithesis-antiphrasis occurs in *Second Corinthians* 3:2–3.

2. I give the most widespread US version; for variants, see Opie and Opie, *The Singing Game*, 220–223.

3. See Rubin, *Memory in Oral Traditions*, esp. 175–193; and Jakobson, "Subliminal Verbal Patterning in Poetry."

4. On selection and combination of linguistic raw materials by the "poetic function," see Jakobson, "Linguistics and Poetics." On avoidance of rhyme, see Quintilian, *Institutio oratoria*, IX. 4. 66–80. For the history of rhyme more generally, see Norden, *Die antike Kunstprosa* 2:811–857.

5. Ebbinghaus, *Über das Gedächtnis*, 68–69; Kittler, *Discourse Networks*, 206–212.

6. Benjamin, *Illuminations*, 224.

7. Lowe, *The Classical Plot and the Invention of Western Narrative*, 110–111 (italics in original).

8. Ibid., 111. These symmetries and correspondences often serve interpreters of the *Iliad* and *Odyssey* as evidence of a single omnipresent architect, namely Homer: see, e.g., Whitman, *Homer and the Heroic Tradition*, and Stanley, *The Shield of Homer*.

9. In "La notion de 'rythme' dans son expression linguistique," Benveniste demonstrated the lexical equivalence of the Greek terms *rhythmos* and *schēma* (both are nominalizations of a verbal root, the former, from *rheō*, meaning a manner of flowing and the latter, from *ekhō*, a manner of being). *Schēma*, with its Latin counterpart *habitus*, is both a constant disposition and an outline of action to be filled out with particulars. Compare Bergson on motor memory: "reminiscences, before they can be activated, need the assistance of a motor element; and . . . demand, before they can be called up, a sort of mental attitude enveloped in a bodily attitude" (*Matière et mémoire*, 265).

10. Cf. Durham, *Coevolution*.

11. Olrik, *Principles for Oral Narrative Research*, 41–52; emphasis added.

12. Olrik limits his generalization to "the European narrative tradition." Anthropologists report many other kinds of device for containing and directing a story. The psychologist Frederick Bartlett subjected a story from the Pacific Northwest (recorded by Boas in *Kathlamet Texts*) to the memory capacities of English university students, and found that they inevitably distorted, or in our terms, reformatted, the tale to fit with their own culturally learned logic (*Remembering*, 118–130).

13. Rubin, *Memory in Oral Traditions*, 90.

14. Mauss, "Les techniques du corps," in *Sociologie et anthropologie*, 367. On this essay's reception, see Bert, *"Les techniques du corps" de Marcel Mauss*. For a return to this essay in light of an extended conception of rhythm, see Michon, *Rythmes, pouvoir, mondialisation*, and "Notes éparses sur le rythme."

15. Ibid., 368.

16. Ibid., 368–369.

17. Ibid., 384.

18. Ibid., 372.

19. Leroi-Gourhan, *Le geste et la parole I: Technique et langage*, 164.

20. Stiegler, *La technique et le temps, 1: La faute d'Épiméthée*, 185. On Stiegler's anthropogenesis, see Fynsk, "Lascaux and the Question of Origins," 6–10.

21. Ibid., 162 (italics in original).

22. Ibid., 171, and see Leroi-Gourhan, *Technique et langage*, 186, 212–223, 298.

23. The opening of McLuhan's *The Gutenberg Galaxy*, with its invocation of Parry and Lord, is symptomatic. The book chronicles "the 'disturbances,' first of literacy, and then of printing" (4): that oral tradition might itself be a disturbance, or an extension of man, is not even imagined.

24. Leroi-Gourhan, *La mémoire et les rythmes*, 103.

25. On rhythm as a means of economizing effort and attention, see Spencer, *Philosophy of Style*, 39–40. But see Bücher, *Arbeit und Rhythmus*, on the arts of dance and music as pleasure-giving forms that grew out of originally functional work rhythms, and thus represent a supplementary value extrinsic to productive effort.

26. Honko, "The *Kalevala* as Performance," in *Theoretical Milestones*, 216. For comparative evidence, see Honko, *Thick Corpus, Organic Variation and Textuality*.

27. Binet, "Les grandes mémoires: Résumé d'une enquête sur les joueurs d'échecs," 829, 841–842, 843.

28. Goody, *The Myth of the Bagre*, 130.

29. Casajus, *L'Aède et le troubadour*, 35–36.

30. *Iliad* 2: 488–491.

31. For "bricolage," see Lévi-Strauss, *La pensée sauvage*, 26–33; for "dream-work," Freud, *The Interpretation of Dreams*, in *Standard Edition*, 4, especially 277–338. On forgetting, see Bartlett, *Remembering*, 200–214, and Freud, *The Psycho-Pathology of Everyday Life*, in *Standard Edition* 6:1–8 (the famous "Signorelli" example).

32. Merleau-Ponty, *Phénoménologie de la perception*, 136 (my translation).

33. Platt, review of Meillet, cited in Parry, *The Making of Homeric Verse*, 9.

34. Merleau-Ponty, *Phénoménologie de la perception*, 140–143. The patient, Schneider, was first described by Kurt Goldstein in *Über die Abhängigkeit der Bewegungen von optischen Vorgängen* (*On the Dependency of Movements on Optical Processes*, 1923).

35. Varela, Thompson and Rosch, *The Embodied Mind*, 9.

36. Maturana and Varela, *Autopoiesis and Cognition*, 78–79. It appears that the brains of Pandits specialized in Vedic recitation, when examined in a scanner, evidence "massive gray matter density and cortical thickness increases in . . . language, memory and visual systems" as well as "differences in hippocampal morphometry matching those previously documented for expert spatial navigators," these differences being explained by "the brain organization implementing formalized oral knowledge systems." See Hartzell et al., "Brains of Verbal Memory Specialists."

37. Drachmann, *Scholia vetera*, 30–31; Eustathius, *Commentarii*, 9.

38. McGann, *The Textual Condition*, 15.

Bibliography

Abraham, Nicolas. *Rythmes de l'oeuvre, de la traduction et de la psychanalyse.* Edited by Nicholas Rand and Maria Torok. Paris: Flammarion, 1985.

Actes du deuxième Congrès International de Linguistes, Genève, 25–29 août 1931. Paris: Maisonneuve, 1933.

Andersen, Øivind. "Oral Tradition." In *Jesus and the Oral Gospel Tradition,* edited by Henry Wansbrough, 17–58. Sheffield: JSOT Press, 1991.

Anderson, Stephen R. *Phonology in the Twentieth Century: Theories of Rules and Theories of Representations.* Chicago: University of Chicago Press, 1985.

André, Charles. "Psychologie experimentale et exégèse: Les conférences du R. P. Jousse à l'Institut Biblique de Rome." *La Croix,* 3 February 1927, 4.

Anonymous (Anon.). "Essay on American Language and Literature." *North-American Review and Miscellaneous Journal* 1 (1815): 307–314.

Antliff, Mark. *Inventing Bergson: Cultural Politics and the Parisian Avant-Garde.* Princeton: Princeton University Press, 1993.

Anzieu, Didier. *Le moi-peau.* Paris: Dunod, 1985.

Archives Nationales, République Française. Ministère de la Culture, database Léonore, dossier number 19800035/58/7098. "Procès-verbal de réception d'un membre de la Légion d'Honneur," number 259.013, dated 11 April 1952, concerning Mademoiselle Etiennette Boucly (1886–1952).

Arend, Walter. *Die typischen Scenen bei Homer.* Berlin: Weidmann, 1933.

Arnold, Claus, and Giacomo Losito. *La censure d'Alfred Loisy (1903): Les documents des Congrégations de l'Index et du Saint-Office.* (Fontes Archivi Sancti Officii Romani, 4.) Vatican City: Libreria Editrice Vaticana, 2009.

———, eds. *'Lamentabili sane exigu' (1907): Les documents préparatoires du Saint Office.* (Fontes Archivi Sancti Officii Romani, 6.) Vatican City: Libreria Editrice Vaticana, 2011.

Arnold, Matthew. *On Translating Homer*. London: Longmans, 1861.

Asher, R. E., and Eugénie J. A. Henderson, eds. *Towards a History of Phonetics*. Edinburgh: Edinburgh University Press, 1981.

Auerbach, Erich. *Mimesis: The Representation of Reality in Western Literature*. Translated by Willard R. Trask. Princeton: Princeton University Press, 1953.

Auroux, Sylvain, ed. *Histoire des idées linguistiques*. 3 vols. Brussels: Madarga, 1989–2000.

Austin, John. *How to Do Things with Words*. Edited by J. O. Urmson and Maria Sbisà. Cambridge, Mass.: Harvard University Press, 1962.

Austin, Norman. *Archery at the Dark of the Moon: Poetic Problems in Homer's Odyssey*. Berkeley: University of California Press, 1975.

Bachelard, Gaston. *L'intuition de l'instant*. Paris: Alcan, 1932.

Bader, Françoise. "Meillet et la poésie indo-européenne." *Cahiers Ferdinand de Saussure* 42 (1988): 97–125.

Baillaud, Bernard. "Sur la manducation de la parole de Jean Paulhan par Marcel Jousse—et réciproquement, ou Par quel bout les prendre." *Nunc* 25 (2011): 67–71.

Bakhtin, Mikhail Mikhailovich. *Problems of Dostoevsky's Poetics*. Translated by Caryl Emerson. Minneapolis: University of Minnesota Press, 1984.

———. "The Problem of Speech Genres." Translated by Vern W. McGee. In *Speech Genres and Other Late Essays*, 60–102. Austin: University of Texas Press, 1986.

Barbe, Noël, and Jean-François Bert, eds. *Penser le concret: André Leroi-Gourhan, André-Georges Haudricourt, Charles Parain*. Paris: Créaphis, 2011.

Barber, Karin. *The Anthropology of Texts, Persons and Publics: Oral and Written Culture in Africa and Beyond*. Cambridge: Cambridge University Press, 2007.

Barclay, John. "Memory Politics: Josephus on Jews in the Memory of the Greeks." In *Memory in the Bible and Antiquity: The Fifth Durham-Tübingen Research Symposium*, edited by Stephen C. Barton, Loren T. Stuckenbruck, and Benjamin G. Wold, 129–142. Tübingen: Mohr Siebeck, 2007.

Baron, Gabrielle. *Marcel Jousse: Introduction à sa vie et à son oeuvre*. Tournai: Casterman, 1965.

———. *Mémoire vivante: Vie et oeuvre de Marcel Jousse*. 2nd edition. Paris: Le Centurion, 1981.

Barthes, Roland. "L'écriture de l'événement." *Communications* 12 (1968): 108–112.

Bartlett, Frederic Charles. *Remembering: A Study in Experimental and Social Psychology*. Cambridge: Cambridge University Press, 1932.

Bastide, Roger. *Art et société*. Paris: Payot, 1977 (1945).

Bateson, Gregory. *Steps to an Ecology of Mind*. New York: Ballantine Books, 1972.

Baudelaire, Charles. *Oeuvres complètes*. Edited by Claude Pichois. 2 vols. Paris: Gallimard, 1975.

Baudet, Jacques. "L'abbé Jean-Pierre Rousselot (1846–1924), ses relations avec la Charente dans sa vie et son oeuvre scientifique." *Bulletins et mémoires de la Société archéologique et historique de la Charente* (1986): 46–68.

Bauman, Richard. "Conceptions of Folklore in the Development of Literary Semiotics." *Semiotica* 39 (1982): 1–20.

Beck, Philippe. "Le geste de la poésie." *Agenda de la pensée contemporaine* 15 (2009): 33–55.

Bell, Alexander Melville. *Visible Speech: The Science of Universal Alphabetics: or Self-Interpreting Physiological Letters, for the Writing of All Languages in One Alphabet.* London: Simpkin, Marshall, 1867.

Benjamin, Walter. *Illuminations: Essays and Reflections.* Edited by Hannah Arendt. Translated by Harry Zohn. New York: Schocken, 1969.

———. *Gesammelte Schriften.* 4 vols. Edited by Rolf Tiedemann and Hermann Schweppenhäuser. Frankfurt am Main: Suhrkamp, 1980.

———. *Gesammelte Schriften.* 6 vols. Edited by Rolf Tiedemann and Hermann Schweppenhäuser. Frankfurt am Main: Suhrkamp, 1991.

———. *Selected Writings.* 4 vols. Edited by Marcus Bullock and Michael W. Jennings. Cambridge, Mass.: Harvard University Press, 1996.

Benoit, Serge, Daniel Blouin, Jean-Yves Dupont, and Gérard Emptoz. "Chronique d'une invention: Le *phonautographe* d'Édouard-Léon Scott de Martinville (1817–1879) et les cercles parisiens de la science et de la technique." *Documents pour l'histoire des techniques* 17 (2009). Available at http://dht.revues.org/502. Accessed 29 July 2013.

Benveniste, Emile. "La notion de 'rythme' dans son expression linguistique." *Journal de psychologie normale et pathologique* 44 (1951): 401–410; reprinted in *Problèmes de linguistique générale*, 327–335. Paris: Gallimard, 1966.

Bérard, Victor. *Un mensonge de la science allemande: les 'Prolégomènes à Homère' de Frédéric-Auguste Wolf.* Paris: Hachette, 1917.

Bergeron, Katherine. "A Bugle, a Bell, a Stroke of the Tongue: Rethinking Music in Modern French Verse." *Representations* 86 (2004): 53–72.

Bergounioux, Gabriel. "Le francien (1815–1914): la linguistique au service de la patrie." *Mots* 19 (1989): 23–40.

Bergson, Henri. *Matière et mémoire: Essai sur la relation du corps à l'esprit.* Paris: Alcan, 1896. Reprint, Paris: Presses Universitaires de France, 1999.

Berliner, Paul. *Thinking in Jazz: The Infinite Art of Improvisation.* Chicago: University of Chicago Press, 1994.

Bernardi, Peter J. *Maurice Blondel, Social Catholicism and Action Française: The Clash over the Church's Role in Society during the Modernist Era.* Washington, D.C.: Catholic University of America Press, 2009.

Bernoville, Gaëtan. "Après la 'Semaine' sur l'unification des Églises." *Les Lettres* 15 (1928): 366–376.

Bersani, Jacques, ed. *Jean Paulhan le souterrain.* Paris: Union générale d'éditions, 1976.

Bert, Jean-François, ed. *"Les techniques du corps" de Marcel Mauss: Dossier critique*. Paris: Publications de la Sorbonne, 2012.

Binet, Alfred. "Les grandes mémoires: Résumé d'une enquête sur les joueurs d'échecs." *Revue des deux mondes* 107 (1893): 826–859.

Black, Matthew. *An Aramaic Approach to the Gospels and Acts*. 3rd edition. Oxford: Oxford University Press, 1967.

Blondel, Maurice. *Carnets intimes*. 2 vols. Paris: Cerf, 1961–66.

Bloomfield, Leonard. *Language*. New York: Henry Holt, 1933.

Boas, Franz, ed. *Kathlamet Texts*. (Bureau of American Ethnology, Bulletin 26.) Washington, D.C.: US Government Printing Office, 1901.

Bobillot, Jean-Pierre. "René Ghil: une mystique matérialiste du langage?" In *René Ghil, De la Poésie-Scientifique et autres écrits*, edited by Jean-Pierre Bobillot, 5–85. Grenoble: ELLUG, 2008.

Bogatyrëv, Petr G. *Voprosy teorii narodnogo iskusstva*. Moscow: Izdatel'stvo iskusstvo, 1971.

———. *Funktionale-strukturale Ethnographie in Europa*. Edited by Klaas-Hinrich Ehlers and Marek Nekula. Heidelberg: Winter, 2011.

Bogatyrëv, Petr G., and Roman Jakobson. "Die Folklore als eine besondere Form des Schaffens." In *Donum Natalicium Schrijnen. Verzameling van Opstellen door Oud-leerlingen en Bevriende Vakgenooten Opgedragen aan Mgr. Prof. Dr. Jos. Schrijnen bij Gelegenheid van zijn Zestigsten Verjaardag, 3 Mei 1929*, 900–913. Nijmegen–Utrecht: Dekker & van de Vegt, 1929. Reprinted in Roman Jakobson, *Selected Writings, vol. 4: Slavic Epic Studies*, 1–15. Translated by John M. O'Hara as "Folklore as a Special Form of Creation," *Folklore Forum* 13 (1980): 1–21, and by Manfred Jacobson as "Folklore as a Special Form of Creativity," in *The Prague School: Selected Writings, 1929–1946*, edited by Peter Steiner, 32–46.

Boissière, Anne. "La part gestuelle du sonore: Expression parlée, expression dansée. Main et narration chez Walter Benjamin." *DEMeter* (2004), available at http://demeter.revue.univ-lille3.fr/manieres/boissiere.pdf. Accessed June 23, 2013.

Bolter, Jay David. *Writing Space: The Computer, Hypertext and the Remediation of Print*. 1991. 2nd edition, Mahwah, N.J.: Lawrence Erlbaum, 2001.

Bolter, Jay David, and Richard Grusin. *Remediation: Understanding New Media*. Cambridge, Mass.: MIT Press, 1999.

Bolton, Thaddeus L. "Rhythm." *American Journal of Psychology* 6 (1894): 145–238.

Bonnefoy, Jehan de (pseudonym of abbé Joseph Brugerette). *Vers l'unité de croyance*. Paris: Nourry, 1907.

Bonnet, Jean-François, and Louis-Jean Boë. "Émergence de la phonétique générale et expérimentale en France au tournant du XXe siècle: de la prise en compte de l'oralité à la recherche d'une transcription 'idéale.'" In *Par monts et par vaux, itinéraires linguistiques et grammaticaux: Mélanges de linguistique générale et française offerts au professeur Martin Riegel pour son soixantième anniversaire par ses collègues et amis*, edited by Claude Buridant, Georges Kleiber, and Jean-Claude Pellat, 43–54. Louvain and Paris: Peeters, 2001.

Boone, Elizabeth Hill, and Walter D. Mignolo, eds. *Writing Without Words: Alternative Literacies in Mesoamerica and the Andes*. Durham, N.C.: Duke University Press, 1994.

Boschian-Campaner, Catherine, ed. *Le vers libre dans tous ses états: Histoire et poétique d'une forme (1886–1914)*. Paris: L'Harmattan, 2009.

Boucly, Étiennette. "Israël auteur classique." *Cahiers juifs* 10 (1934): 34–39.

———. "Le style oral chez les Rabbis d'Israël." *Cahiers juifs* 7 (1934): 62–66.

———. "Le style oral dans les milieux palestiniens: L'oeuvre du Père Marcel Jousse." *La Revue juive de Genève* 3 (May 1934): 331–336.

———. "Le style oral palestinien et sa présentation typographique." *Cahiers juifs* 8 (1934): 187–188.

———. "La mimique hébraïque et la rythmo-pédagogie vivante." *Cahiers juifs* 15 (1935): 199–210.

Brain, Robert. "Genealogy of 'Zang Tumb Tumb': Experimental Phonetics, Vers Libre, and Modernist Sound Art." *Grey Room* 43 (2011): 88–117.

———. "Representation on the Line: Graphic Recording Instruments and Scientific Modernism." In Clarke and Henderson, *From Energy to Information*, 151–212.

———. "Standards and Semiotics." In Lenoir, *Inscribing Science*, 249–284.

———. "Materialising the Medium: Ectoplasm and the Quest for Supra-Normal Biology in Fin-de-Siècle Science and Art." In Enns and Trower, *Vibratory Modernism*, 115–144.

Brain, Robert, and W. Norton Wise. "Muscles and Engines: Indicator Diagrams and Helmholtz's Graphical Methods." In *Universalgenie Helmholtz. Rückblick nach 100 Jahren*, edited by Lorenz Krüger, 124–148. Berlin: Akademie Verlag, 1994.

Brathwaite, Edward Kamau. *History of the Voice: The Development of Nation Language in Anglophone Caribbean Poetry*. London: New Beacon Books, 1984.

Bravo, Benedetto. "*Critice* in the Sixteenth and Seventeenth Centuries and the Rise of a Notion of Historical Criticism." In *History of Scholarship: A Selection of Papers from the Seminar on the History of Scholarship Held Annually at the Warburg Institute*, edited by Christopher Ligota and Jean-Louis Quantin, 135–196. Oxford: Oxford University Press, 2006.

Bréal, Michel. "Des lois phoniques: À propos de la création du laboratoire de phonétique expérimentale au Collège de France." *Mémoires de la Société de linguistique de Paris* 10 (1898): 1–11.

Bremer, Jan M., I. J. F. de Jong, and J. Kalff, eds. *Homer: Beyond Oral Poetry—Recent Trends in Homeric Interpretation*. Amsterdam: Grüner, 1987.

Bremond, Henri. *La poésie pure*. Paris: Grasset, 1926.

———. *Prière et poésie*. Paris: Grasset, 1926.

Bremond, Henri, and Maurice Blondel. *Correspondance*. 3 vols. Edited by André Blanchet. Paris: Aubier-Montaigne, 1971.

Briand, Henri. "Une conférence du Professeur Eugen Fischer à Paris." *Revue anthropologique* 51 (1941): 195–196.

Brunot, Ferdinand. "L'inscription de la parole." *La Nature* 998 (6 July 1892): 97–98.

Bücher, Karl. *Arbeit und Rhythmus.* 3rd edition. Leipzig: Teubner, 1902.

Bultmann, Rudolf. *History of the Synoptic Tradition.* Translated by John Marsh. Oxford: Blackwell, 1963.

Burns, Christy L. *Gestural Politics: Stereotype and Parody in Joyce.* Albany: State University of New York Press, 2000.

Busse Berger, Anna Maria. "L'invention du temps mesuré au XIIIe siècle." In *Les Écritures du temps (musique, rythme, etc.),* edited by Fabien Lévy, 19–54. Paris: L'Harmattan, 2001.

———. *Medieval Music and the Art of Memory.* Berkeley: University of California Press, 2005.

Büttgen, Philippe. "Lessing et la question du prêche: Philosophie, théologie, pastorat." *Les études philosophiques* 65 (2003): 213–243.

Cadore, Henri de. "*L'Ouest-Éclair* et les deux Ralliements (1899–1930): Contribution à l'intégration des Catholiques de l'Ouest dans la République." In Lagrée et al., *L'Ouest-Éclair,* 55–78.

Caesar, Gaius Julius. *Commentarii de Bello Gallico.* Edited by Heinrich Meusel. 3 vols. Berlin: Weidmann, 1961.

Callimachus. Edited by Rudolf Pfeiffer. 2 vols. Oxford: Oxford University Press, 1949, 1953.

Campbell, Timothy C. *Wireless Writing in the Age of Marconi.* Minneapolis: University of Minnesota Press, 2006.

Candidus, Petrus (or Decembrius). *Homerokentra, hoc est centones ex Homero.* Venice, 1502.

Cerquiglini, Bernard. *Éloge de la variante: Histoire critique de la philologie.* Paris: Seuil, 1989.

———. *Une langue orpheline.* Paris: Minuit, 2007.

Certeau, Michel de. "Pour une nouvelle culture: prendre la parole." *Études* 329 (1968): 29–42.

———. "Une anthropologie du geste: Marcel Jousse." *Études* 332 (1970): 770–773.

———. *La prise de parole et autres écrits politiques,* edited by Luce Giard. Paris: Seuil, 1994.

Certeau, Michel de, Dominique Julia, and Jacques Revel. *Une politique de la langue. La Révolution française et les patois: l'enquête de Grégoire.* Paris: Gallimard, 2002.

Charcot, Jean-Martin. *Leçons sur les maladies du système nerveux, tome 1.* Paris: Delahaye et Lecrosnier, 1880.

Chasles, Emile. *Michel de Cervantes: Sa vie, son temps, son oeuvre politique et littéraire.* Paris: Didier, 1866.

Chen Kaige. *Huang tudi* (Yellow Earth). Nanning: Guangxi Film Studios, 1984.

Chevalier, Jean-Claude. "Le prophète et le roi: Tourtoulon devant G. Paris." In *Et multum et multa: Festschrift für Peter Wunderli zum 60. Geburtstag*, edited by Edeltraud Werner, 45–55. Tübingen: Narr, 1998.

Chomsky, Noam. *Syntactic Structures*. The Hague: Mouton, 1957.

Claparède, Edouard. "Existe-t-il des images verbo-motrices?" *Archives de psychologie* 13 (1923): 93–103.

Clarke, Bruce, and Linda Dalrymple Henderson, eds. *From Energy to Information: Representation in Science and Art, Technology and Literature*. Stanford: Stanford University Press, 2002.

Clarke, Edwin, and L. S. Jacyna. *Nineteenth-Century Origins of Neuroscientific Concepts*. Berkeley: University of California Press, 1987.

Colin, Pierre. *L'audace et le soupçon. La crise du modernisme dans le catholicisme français (1893–1914)*. Paris: Desclée de Brouwer, 1997.

Colmet, Pierre (=abbé Paul Boulin). "L'abolition des 'Amis d'Israël.'" *R.I.S.S. (Revue Internationale des sociétés secrètes)* 18 (23 April 1928): 369–386.

Colum, Mary, and Padraic Colum. *Our Friend James Joyce*. New York: Doubleday, 1958.

Compagnon, Antoine. *Les antimodernes: De Joseph de Maistre à Roland Barthes*. Paris: Gallimard, 2005.

Conklin, Alice L. *In the Museum of Man: Race, Anthropology and Empire in France, 1850–1950*. Ithaca, N.Y.: Cornell University Press, 2013.

Connelly, John. *From Enemy to Brother: The Revolution in Catholic Teaching on the Jews, 1933–1965*. Cambridge, Mass.: Harvard University Press, 2012.

Connerton, Paul. *How Societies Remember*. Cambridge: Cambridge University Press, 1989.

Coustille, Charles. "Pour une soutenance de thèse de Jean Paulhan." Available at www.fabula.org/colloques/document1702.php. Accessed 15 July 2013.

Crane, Gregory, ed. *Perseus Digital Library*. www.perseus.tufts.edu.

Crielaard, Jan Paul, ed. *Homeric Questions*. Amsterdam: Gieben, 1995.

Cros, Charles. *Oeuvres complètes*. Edited by Louis Forestier and Pierre-Olivier Walzer. Paris: Gallimard, 1970.

Culbert, John. *Paralyses: Literature, Travel and Ethnography in French Modernity*. Lincoln: University of Nebraska Press, 2010.

Cushing, Frank Hamilton. "Manual Concepts: A Study of the Influence of Hand-Usage on Culture-Growth." *American Anthropologist* 5 (1892): 289–317.

Dacier, Anne Le Favre. *Des causes de la corruption du goût*. Paris: 1714.

Dagognet, François. *Étienne-Jules Marey: La passion de la trace*. Paris: Hazan, 1987.

Daniels, Peter T., and William Bright. *The World's Writing Systems*. New York: Oxford University Press, 1996.

Darwin, Charles. *The Expression of the Emotions in Man and Animals*. New York: Appleton, n.d.

d'Aubignac, Hédelin, François, abbé. *Conjectures académiques ou Dissertation sur l'Iliade [1715]*. Edited by Victor Magnien. Paris: Hachette, 1925.

Davidson, Donald. *Inquiries into Truth and Interpretation*. Oxford: Clarendon Press, 2001.

Deffayet-Loupiac, Laurence. "Le rôle du Père Joseph Bonsirven dans le renouveau du dialogue judéo-chrétien." *Revue d'histoire de l'église de France* 222 (2002): 81–103.

———. "Le Père Joseph Bonsirven: Un parcours fait d'ombres et de lumières." *Archives juives* 40 (2007): 30–44.

DeJean, Joan. *Ancients Against Moderns: Culture Wars and the Making of a Fin de Siècle*. Chicago: University of Chicago Press, 1997.

Delbreil, Jean-Claude. "*L'Ouest-Éclair* et le parti démocrate populaire." In Lagrée et al., *L'Ouest-Éclair*, 79–100.

Denker, Alfred, and Holger Zaborowski, eds. *Heidegger und der Nationalsozialismus, I: Dokumente*. Freiburg: Alber, 2009.

Denzinger, Heinrich, and Adolph Schönmetzer, eds. *Enchiridion symbolorum, definitionum et declarationum de rebus fidei et morum*. 32nd edition. Barcelona: Herder, 1962.

Deroy, Louis. "Le nom d'Homère." *L'antiquité classique* 41 (1972): 427–439.

Derrida, Jacques. *De la grammatologie*. Paris: Minuit, 1967.

———. *L'écriture et la différence*. Paris: Seuil, 1967.

———. *Marges de la philosophie*. Paris: Minuit, 1972.

———. *La dissémination*. Paris: Seuil, 1974.

———. *La carte postale de Socrate à Freud et au-delà*. Paris: Aubier-Flammarion, 1980.

Desmet, Piet, Peter Lauwers, and Pierre Swiggers. "Le développement de la dialectologie française avant et après Gilliéron." In *Géographie linguistique et biologie du langage: autour de Jules Gilliéron*, edited by Peter Lauwers, Marie-Rose Simoni-Aurembou, and Pierre Swiggers, 17–64. Leuven: Peeters, 2002.

de Souza, Robert. *Où nous en sommes: La victoire du silence*. Paris: Floury, 1906.

———. *Du rythme en français*. Paris: Welter, 1912.

———. "La phonétique expérimentale et son créateur l'Abbé Rousselot." *L'Illustration* 4219 (12 Jan. 1924): 37–39.

de Vet, Thérèse. "Parry in Paris: Structuralism, Historical Linguistics, and the Oral Theory." *Classical Antiquity* 24 (2005): 257–284.

Dewey, John. "The Reflex Arc Concept in Psychology." *Psychological Review* 3 (1896): 357–370.

Dictionnaire de l'Académie française. 7th edition. Paris: Firmin Didot, 1877.

Dominicy, Marc. "La formule chez Milman Parry: une étude épistémologique." Manuscript, 2000. Available at http://difusion.academiewb.be/vufind/Record/ULB-DIPOT:oai:dipot.ulb.ac.be:2013/129764/Holdings. Accessed 24 July 2013.

———. *L'évocation poétique*. Paris: Garnier, 2011.

Donatelli, John M. P. "'To Hear with Eyes': Orality, Print Culture, and the Textuality of Ballads." In *Ballads and Boundaries: Narrative Singing in an*

Intercultural Context, edited by James Porter, 347–357. Los Angeles: Department of Ethnomusicology and Systematic Musicology, 1995.

Dow, James R. "Hans Naumann's *gesunkenes Kulturgut* and *primitive Gemeinschaftskultur.*" *Journal of Folklore Research* 51 (2014): 49–100.

Drachmann, A. B., ed. *Scholia vetera in Pindari carmina*. 3 vols. Leipzig: Teubner, 1903–1927.

d'Udine, Jean (pseudonym of Albert Cozanet). *L'art et le geste*. Paris: Alcan, 1910.

Dué, Casey, and Mary Ebbott. "Digital Criticism: Editorial Standards for the Homer Multitext." *Digital Humanities Quarterly* 3 (2009). Available at www.digitalhumanities.org/dhq/vol/3/1/000029/000029.html. Accessed 24 June 2014.

Dugas-Montbel, Jean-Baptiste. *Observations sur l'Odyssée d'Homère*. 2 vols. Paris: Didot, 1830.

Dujardin, Jean. "Les relations entre chrétiens et juifs depuis 50 ans: Aperçu historique." *Théologiques* 11 (2003): 17–33.

Dumas, Georges, ed. *Traité de psychologie*. 2 vols. Paris: Alcan, 1923–24.

———, ed. *Nouveau traité de psychologie*. 7 vols. Paris: Alcan and Presses Universitaires de France, 1930–1949.

Dumézil, Georges. "La tradition druidique et l'écriture: le vivant et le mort." In *Pour un temps: Georges Dumézil*, edited by Jacques Bonnet, 325–338. (Cahiers pour un temps, 3.) Paris: Centre Georges Pompidou/Editions Pandora, 1981.

Dupuy, Jean-Pierre. *Aux origines des sciences cognitives*. Paris: La Découverte, 1994. Translated by M. B. DeBevoise as *The Mechanization of the Mind: On the Origins of Cognitive Science*. Princeton: Princeton University Press, 2000.

Durham, William. *Coevolution: Genes, Culture, and Human Diversity*. Stanford: Stanford University Press, 1991.

Durkheim, Emile. *Les Formes élémentaires de la vie religieuse*. 1911. Reprint, Paris: Presses Universitaires de France, 1998.

Durkheim, Emile, and Marcel Mauss. "De quelques formes primitives de classification." *L'année sociologique* 6 (1901–1902): 1–72.

Ebbinghaus, Hermann. *Über das Gedächtnis: Untersuchungen zur experimentellen Psychologie*. Leipzig: Duncker & Humboldt, 1885.

Eco, Umberto. *Foucault's Pendulum*. Translated by William Weaver. New York: Harcourt Brace, 1989.

Edwards, Mark W. "Homer and Oral Tradition: The Formula, Part I." *Oral Tradition* 1 (1986): 171–230.

———. "Homer and Oral Tradition: The Formula, Part II." *Oral Tradition* 3 (1988): 11–60.

———. "Homer and Oral Tradition: The Type-Scene." *Oral Tradition* 7 (1992): 284–330.

———. "Topos and Transformation in Homer." In Bremer, de Jong, and Kalff, *Homer: Beyond Oral Poetry*, 47–60.

Eikhenbaum, Boris. "The Illusion of *Skaz*." Translated by Martin P. Rice. *Russian Literature Triquarterly* 12 (1975): 233–236.

Eisenstein, Elizabeth L. *The Printing Press as an Agent of Change: Communications and Cultural Transformations in Early-Modern Europe.* Cambridge: Cambridge University Press, 1979.

Eliade, Mircea. "Littérature orale." In *Histoire des littératures, 1: Littératures anciennes, orientales et orales*, edited by Raymond Queneau, 3–26. (Encyclopédie de la Pléiade, 1.) Paris: Gallimard, 1956.

Ellis, Alexander John. *The Alphabet of Nature.* Bath: Pitman, 1845.

———. *A Plea for Phonotypy and Phonography; Or, Speech-Printing and Speech-Writing.* Bath: Pitman, 1845.

Enns, Anthony, and Shelley Trower, eds. *Vibratory Modernism.* New York: Palgrave, 2013.

Erbse, Hartmut, ed. *Scholia Graeca in Homeri Iliadem (scholia vetera).* Berlin: de Gruyter, 1971.

Erlich, Victor. *Russian Formalism: History—Doctrine.* The Hague: Mouton, 1955.

———. *Child of a Turbulent Century.* Evanston, Ill.: Northwestern University Press, 2006.

Ernst, Ulrich, and Peter-Erich Neuser, eds. *Die Genese der europäischen Endreimdichtung.* (*Wege der Forschung*, vol. 444.) Darmstadt: Wissenschaftliche Buchgesellschaft, 1977.

Falkenhausen, Lothar von. *Suspended Music: The Chime-Bells of the Chinese Bronze Age.* Berkeley: University of California Press, 1993.

Farmer, William R. *The Synoptic Problem.* 2nd edition. Dillsboro, N.C.: Western North Carolina Press, 1976.

Farrell, Thomas J. "Early Christian Creeds and Controversies in the Light of the Orality-Literacy Hypothesis." *Oral Tradition* 2 (1987): 139–149.

Faublée, Jacques. "Jean Paulhan malgachisant." *Journal de la Société des Africanistes* 40 (1970): 151–159.

———. "Les études littéraires malgaches de Jean Paulhan." *Journal de la Société des Africanistes* 54 (1984): 79–93.

Faure, Christian. *Le projet culturel de Vichy: Folklore et révolution nationale, 1940–1944.* Lyon: Presses Universitaires de Lyon, 1989.

Feaster, Patrick. *Pictures of Sound: One Thousand Years of Educed Audio, 980–1980.* Atlanta: Dust-to-Digital, 2012.

Fénelon, François de Salignac de la Mothe. *Démonstration de l'existence de Dieu, tirée de la connaissance et proportionnée à la faible intelligence des plus simples.* Paris, 1713.

Fenik, Bernard. *Typical Battle Scenes in the Iliad: Studies in the Narrative Techniques of Homeric Battle Description.* (Hermes Einzelschriften, 21.) Wiesbaden: Steiner, 1968.

———, ed. *Homer: Tradition and Invention.* (Cincinnati Classical Studies, n.s. 2.) Leiden: Brill, 1978.

Ferraris, Maurizio. *Documentality: Why It Is Necessary to Leave Traces.* New York: Fordham University Press, 2013.

Fhima, Catherine. "Aux sources d'un renouveau identitaire juif en France: André Spire et Edmond Fleg." *Mil neuf cent* 13 (1995): 171–189.

Ficino, Marsilio. *Commentaries on Plato, vol. 1: Phaedrus and Ion.* Trans. Michael J. B. Allen. Cambridge, Mass.: Harvard University Press, 2008.

Finnegan, Ruth. *Oral Poetry: Its Nature, Significance and Social Context.* Cambridge: Cambridge University Press, 1977.

Flavius Josephus. *The Genuine Works of Flavius Josephus [Translated] by the Late William Whiston, M.A.* Edited by Samuel Burder. 4 vols. New York: Borradaile, 1824.

———. *Flavii Iosephi Iudaei opera omnia.* Edited by Karl-Ernst Richter. 6 vols. Leipzig: Schwickert, 1826.

———. *Against Apion*, tr. and comm. John M. G. Barclay. (*Flavius Josephus, Translation and Commentary*, ed. Steve Mason, vol. 10.) Leiden: Brill, 2007.

Foley, John Miles. *The Theory of Oral Composition: History and Methodology.* Bloomington: Indiana University Press, 1988.

———. *Oral-Formulaic Theory: A Folklore Casebook.* New York: Garland, 1990.

———. *The Singer of Tales in Performance.* Bloomington: Indiana University Press, 1995.

———. "The Impossibility of Canon." In Foley, *Teaching Oral Traditions*, 13–33.

———, ed. *Oral Traditional Literature: A Festschrift for Albert Bates Lord.* Columbus, Ohio: Slavica, 1981.

———, ed. *Teaching Oral Traditions.* New York: Modern Language Association, 1998.

Fontaine, Julien. *Les infiltrations kantiennes et protestantes et le clergé français: Études complémentaires.* Paris: Retaux, 1902.

———. *La synthèse du modernisme.* Paris: Lethielleux, 1912.

Ford, Andrew. "The Classical Definition of ΡΑΨΩΙΔΙΑ." *Classical Philology* 83 (1988): 300–307.

———. *Homer: The Poetry of the Past.* Ithaca, N.Y.: Cornell University Press, 1992.

Fouilloux, Etienne. *Une Église en quête de liberté: La pensée catholique française entre modernisme et Vatican II, 1914–1962.* Paris: Desclée de Brouwer, 1998.

Fraisse, Pierre. "Contribution à l'étude du rythme en tant que forme temporelle." *Journal de psychologie normale et pathologique* 39 (1946): 283–304.

Freud, Sigmund. *Standard Edition of the Complete Psychological Works.* Edited by James Strachey. London: Hogarth Press, 1953–74.

Frey, Jean-Baptiste. "Le Pater est-il juif ou chrétien?" *Revue Biblique* 24 (1915): 556–563.

Fromont, Marie-Françoise. *L'enfant mimeur: L'anthropologie de Marcel Jousse et la pédagogie.* Paris: Épi, 1978.

Fynsk, Christopher. "Lascaux and the Question of Origins." *Poiesis* 5 (2003): 6–19.

Gacon, François. *Homère vengé ou Réponse à M. de la Motte sur l'Iliade*. Paris: Ganeau, 1715.

García, John F. "Milman Parry and A. F. Kroeber: Americanist Anthropology and the Oral Homer." *Oral Tradition* 16 (2001): 58–84.

Garrabé, Jean. "Martial, ou Pierre Janet et Raymond Roussel." *Annales Médico-Psychologiques* 166 (2008): 225–231.

Garron, Isabelle. "La part typographique: Pour une anthropologie de la page imprimée. Premières balises." *Communication et langages* 134 (2002): 59–74.

Gasparov, M. L. *A History of European Versification*. Translated by G. S. Smith and Marina Tarlinskaja. Oxford: Clarendon Press, 1996.

Gates, Henry Louis, Jr. *The Signifying Monkey*. New York: Oxford University Press, 1987.

Gauchat, Louis. "L'unité phonétique dans le patois d'une commune." In *Aus romanischen Sprachen und Literaturen: Festchrift Heinrich Morf*, 175–232. Halle: Niemeyer, 1905.

Gauchet, Marcel. *L'inconscient cérébral*. Paris: Seuil, 1992.

Gelb, I. J. *A Study of Writing*. Chicago: University of Chicago Press, 1963.

Geller, Jay. *The Other Jewish Question: Identifying the Jew and Making Sense of Modernity*. New York: Fordham University Press, 2011.

Genette, Gérard. *Mimologiques: Voyage en Cratylie*. Paris: Seuil, 1976.

Gerhardsson, Birger. *Memory and Manuscript: Oral Tradition and Written Transmission in Rabbinic Judaism and Early Christianity*. 1961; reprint, Grand Rapids, Mich.: Eerdmans, 1998.

Gesemann, Gerhard. *Studien zur Südslavischen Volksepik* (1926). In *Gesammelte Abhandlungen*, 1:235–339. Neuried: Hieronymus, 1981.

Ghesquier-Pourcin, Danièle, ed., *Énergie, science et philosophie au tournant des XIXe et XXe siècles*. 2 vols. Paris: Hermann, 2010.

Gibert, Pierre. *L'invention critique de la Bible: XVe–XVIIIe siècle*. Paris: Gallimard, 2010.

Gibson, James J. *The Senses Considered as Perceptual Systems*. Boston: Houghton Mifflin, 1966.

Gieseler, Johann Karl Ludwig. *Historisch-kritischer Versuch über die Entstehung und die frühesten Schicksale der schriftlichen Evangelien*. Leipzig: Engelmann, 1818.

Gilliéron, Jules. *Les étymologies des étymologistes et celles du peuple*. Paris: Champion, 1922.

Gitelman, Lisa. *Scripts, Grooves, and Writing Machines: Representing Technology in the Edison Era*. Stanford: Stanford University Press, 1999.

———. *Always Already New: Media, History, and the Data of Culture*. Cambridge, Mass.: MIT Press, 2006.

Gitelman, Lisa, and Geoffrey B. Pingree, *New Media, 1740–1915*. Cambridge, Mass.: MIT Press, 2003.

Godzich, Wlad, and Jeffrey Kittay. *The Emergence of Prose: An Essay in Prosaics*. Minneapolis: University of Minnesota Press, 1987.

Goichot, Emile. *Alfred Loisy et ses amis*. Paris: Cerf, 2002.

Golston, Robert. *Rhythm and Race in Modernist Poetry and Science.* New York: Columbia University Press, 2008.

Gomme, A. W. *An Historical Commentary on Thucydides.* 5 vols. Oxford: Oxford University Press, 1945–1981.

Goodman, Nelson. *Languages of Art.* Indianapolis: Bobbs-Merrill, 1968.

Goody, Jack. *The Myth of the Bagre.* Oxford: Clarendon Press, 1972.

———. *The Domestication of the Savage Mind.* Cambridge: Cambridge University Press, 1977.

———. *The Logic of Writing and the Organization of Society.* Cambridge: Cambridge University Press, 1986.

———. *The Interface Between the Written and the Oral.* Cambridge: Cambridge University Press, 1987.

Goody, Jack, and Ian Watt. "The Consequences of Literacy." *Comparative Studies in History and Society* 5 (1963): 304–345; also in *Literacy in Traditional Societies,* edited by Jack Goody, 27–68. Cambridge: Cambridge University Press, 1968.

Goody, Jack, and S. W. D. K. Gandah, eds. *Une récitation du Bagré.* Paris: Armand Colin, 1980.

Gould, Glenn. *The Glenn Gould Reader.* Edited by Tim Page. Toronto: Key Porter Books, 1998.

Grafton, Anthony. *Forgers and Critics: Creativity and Duplicity in Western Scholarship.* Princeton: Princeton University Press, 1990.

Grammont, Maurice. *Traité de phonétique.* Paris: Delagrave, 1933.

Grandmaison, Léonce de. "Le style oral: En marge d'un mémoire de psychologie linguistique." *Études* 62 (1925): 685–705.

———. "L'odyssée spirituelle d'un Moderniste: du Christianisme au Judaïsme." *Études* 64 (1927): 641–660.

———. *Jésus-Christ, sa personne, son message, ses preuves.* 2 vols. Paris: Beauchesne, 1928.

Granet, Marcel. *Fêtes et chansons anciennes de la Chine.* 1919. 2nd edition, Paris: Leroux, 1929. Translated by E. D. Edwards as *Festivals and Songs of Ancient China.* London: Routledge and Kegan Paul, 1932.

———. "Quelques particularités de la langue et de la pensée chinoises." *Revue philosophique* 89–90 (1920): 98–128, 161–195.

———. *La pensée chinoise.* 1934; reprint, Paris: Albin Michel, 1999.

Griffin, Jasper. "Homer and Excess." In Bremer, de Jong, and Kalff, *Homer: Beyond Oral Poetry,* 85–104.

Griswold, Charles L., Jr. *Self-Knowledge in Plato's Phaedrus.* 2nd edition. University Park: Pennsylvania State University Press, 2010.

Guérinel, Rémy. "Dans la succession de l'Abbé Rousselot: Marcel Jousse, SJ (1886–1961)." In *Sur les pas de Marey: Science(s) et cinéma,* edited by Thierry Lefebvre and Jacques Malthête, 243–251. Paris, L'Harmattan, 2004.

———. "Déchiffrer l'énigme Marcel Jousse (1886–1961) au regard de l'éclipse de Pierre Janet (1859–1947)." *Janetian Studies* 3 (2006): 1–12.

———. "Marcel Jousse entre Pierre Janet et Joseph Morlaas." *Annales Médico-psychologiques* 166 (2008): 232–237.

Guillaumont, Antoine. Review of Gabrielle Baron, "*Marcel Jousse. Introduction à sa vie et à son oeuvre*". *Revue de l'histoire des religions* 175 (1969): 236–237.

Guillory, John. "Genesis of the Media Concept." *Critical Inquiry* 36 (2010): 321–362.

Gumbrecht, Hans Ulrich. "Rhythm and Meaning." In *Materialities of Communication*, edited by Hans Ulrich Gumbrecht and K. Ludwig Pfeiffer, 170–182. Stanford: Stanford University Press, 1994.

Guo Qingfan, ed., *Zhuangzi jishi*. Taipei: Hanjing, 1982.

Güttgemans, Erhardt. "Fundamentals of a Grammar of Oral Literature." In Jason and Segal, *Patterns in Oral Literature*, 77–97.

Guyau, Jean-Marie. "La mémoire et le phonographe." *Revue philosophique de la France et de l'étranger* 9 (1880): 319–322.

Haas, Christina. *Writing Technology: Studies on the Materiality of Literacy*. Mahwah, N.J.: Lawrence Erlbaum Associates, 1996.

Hainsworth, J. B. *The Flexibility of the Homeric Formula*. Oxford: Oxford University Press, 1968.

———. "The Criticism of an Oral Homer." *JHS* 90 (1970): 90–98.

———. "Good and Bad Formulae." In *Homer: Tradition and Invention*, edited by Bernard C. Fenik, 41–50. Leiden: Brill, 1978.

Halflants, Paul. "Le sémitisme chez les 'Ecrivains Catholiques.'" *La Revue catholique des idées et des faits*, 30 December 1927, 13.

Hall, G. Stanley, and Joseph Jastrow. "Studies of Rhythm." *Mind* 11 (1886): 55–62.

Hall, Jason David. "Mechanized Metrics: From Verse Science to Laboratory Prosody, 1880–1918." *Configurations* 17 (2009): 285–308.

Hanse, Olivier. *À l'école du rythme. Utopies communautaires allemandes autour de 1900*. Saint-Étienne: Publications de l'Université de Saint-Étienne, 2010.

Hardcastle, William J., and John Laver, eds. *The Handbook of Phonetic Sciences*. Malden, Mass.: Blackwell, 1999.

Hardin, Garrett. "The Tragedy of the Commons." *Science*, n.s. 162 (1968): 1243–1248.

Harnack, Adolf. *Das Wesen des Christentums: Sechzehn Vorlesungen vor Studierenden aller Facultäten im Wintersemester 1899/1900 an der Universität Berlin gehalten*. Leipzig: Hinrichs, 1902. Translated by Thomas Bailey Saunders as *What is Christianity? Lectures Delivered in the University of Berlin during the Winter-Term 1899–1900*. New York: Putnam, 1903.

———. "Über einige Worte Jesu, die nicht in den kanonischen Evangelien stehen, nebst einen Anhang über die ursprüngliche Inhalt des Vater-unsers." *Sitzungsberichte der königlichen preussischen Akademie der Wissenschaften* 16 (1904): 170–208.

Hartzell, James F., et al. "Brains of Verbal Memory Specialists Show Anatomical Differences in Language, Memory and Visual Systems." *NeuroImage* (2015), in press.

Haugen, Kristine Louise. "Ossian and the Invention of Textual History." *Journal of the History of Ideas* 59 (1998): 309–327.

Havelock, Eric A. *Preface to Plato*. Cambridge, Mass.: Harvard University Press, 1963.

Hayles, Katherine. *Writing Machines*. Cambridge, Mass.: MIT Press, 2002.

Hayot, Eric, and Edward Wesp. "Style, Strategy and Mimesis in Ergodic Literature." *Comparative Literature Studies* 41 (2004): 404–423.

Heath, Stephen. "Ambiviolences: Notes for Reading Joyce." In *Post-Structuralist Joyce: Essays from the French*, edited by Derek Attridge and Daniel Ferrer, 31–68. Cambridge: Cambridge University Press, 1984.

Hecht, Jennifer Michael. *The End of the Soul: Scientific Modernity, Atheism, and Anthropology in France*. New York: Columbia University Press, 2003.

Hédelin, François, abbé d'Aubignac. *Conjectures académiques, ou Dissertation sur l'Iliade*. Edited by Gérard Lambin. Paris: Honoré Champion, 2010.

Hegel, Georg Wilhelm Friedrich. *Hegels theologische Jugendschriften*. Edited by Hermann Nohl. Tübingen: Mohr, 1907.

Helmholtz, Hermann. *On the Sensations of Tone*. Translated by Alexander J. Ellis. New York: Dover, 1954 (1862).

Herder, Johann Gottfried. *Von Gottes Sohn, der Welt Heiland, nach Johannes Evangelium*. Riga: Hartknoch, 1797.

———. *Werke*, ed. Karl-Gustav Gerold. 2 vols. Munich: Hanser, 1953.

Heubeck, Alfred. *Schrift*. Vol. 3, chapter 10 of Friedrich Matz and Hans-Günter Buchholz, eds., *Archaeologia Homerica: Die Denkmäler und das frühgriechische Epos*. Göttingen: Vandenhoeck and Ruprecht, 1979.

Hoff, H. E., and Geddes, L. A. "Graphic Registration Before Ludwig: The Antecedents of the Kymograph." *Isis* 50 (1959): 5–21.

———. "The Beginnings of Graphic Recording." *Isis* 53 (1962): 287–324.

Hollier, Denis, ed. *Le Collège de Sociologie, 1937–1939*. 2nd edition. Paris: Gallimard, 1995. Translated as *The College of Sociology 1937–39*. Minneapolis: University of Minnesota Press, 1988.

Holoka, James P. "Homeric Originality: A Survey." *Classical World* 66 (1972–73): 257–293.

Homer. *Opera*. Edited by David B. Monro and Thomas W. Allen. 5 vols. Oxford: Clarendon Press, 1975.

Honko, Lauri, ed. *Thick Corpus, Organic Variation and Textuality in Oral Tradition*. Helsinki: Finnish Literature Society, 2000.

———. *Theoretical Milestones: Selected Writings of Lauri Honko*. Edited by Pekka Hakamies and Anneli Honko. Helsinki: Academia Scientiarum Fennica, 2013.

———, ed. *Textualization of Oral Epics*. New York: Mouton–de Gruyter, 2000.

Houdard de la Motte, Antoine. *L'Iliade, poème, avec un discours sur Homère*. Paris: Dupuis, 1714.

Houtin, Albert. *La question biblique chez les catholiques de France au XIXe siècle*. Paris: Picard, 1902.

Hung, Chang-tai. *Going to the People: Chinese Intellectuals and Folk Literature.* Cambridge, Mass.: Harvard University Asia Center, 1986.

Hunter, Ian M. "Lengthy Verbal Recall: The Role of Text." *Progress in the Psychology of Language* 1 (1985): 207–235.

Illich, Ivan. *In the Vineyard of the Text: A Commentary to Hugh's Didascalion.* Chicago: University of Chicago Press, 1993.

Institut International d'Anthropologie. *IIIe session, Amsterdam, 20–29 septembre 1927.* Paris: Nourry, 1928.

Jacquin, Robert. *Notions sur le langage d'après les travaux du P. Marcel Jousse.* (Programme à option du Baccalauréat de Philosophie.) Paris: Vrin, 1929.

Jakobson, Roman. *Selected Writings,* vol. 1: *Phonological Studies.* The Hague: Mouton, 1966.

———. *Selected Writings,* vol. 4: *Slavic Epic Studies.* The Hague: Mouton, 1966.

———. *Language in Literature.* Edited by Krystyna Pomorska and Stephen Rudy. Cambridge, Mass.: Harvard University Press, 1987.

———. *On Language.* Edited by Linda R. Waugh and Monique Monville-Burston. Cambridge, Mass.: Harvard University Press, 1990.

———. *My Futurist Years.* Edited by Bengt Jangfelt; translated by Stephen Rudy. New York: Marsilio, 1997.

———. "Closing Statement: Linguistics and Poetics." In Sebeok, *Style in Language,* 350–377.

———. "Subliminal Verbal Patterning in Poetry." *Poetics Today* 2 (1980): 127–136.

Jakobson, Roman, and Petr Bogatyrev. "K probleme razmezhevaniya fol'kloristiki i literaturovedeniya." In Jakobson, *Selected Writings,* vol. 4: *Slavic Epic Studies,* 16–18. Translated by Herbert Eagle as "On the Boundary between Studies of Folklore and Literature," in *Readings in Russian Poetics: Formalist and Structuralist Views,* edited by Ladislav Matejka and Krystyna Pomorska, 91–93. Cambridge, Mass.: MIT Press, 1971.

James, William. "What Is an Emotion?" *Mind* 9 (1884): 188–205.

Janet, Pierre. *L'automatisme psychologique: Essai de psychologie expérimentale sur les formes inférieures de l'activité humaine.* Paris: Editions Odile Jacob, 1998 (1889).

———. *De l'angoisse à l'extase: études sur les croyances et les sentiments.* 2 vols. Paris: Alcan, 1926–28.

Janet, Pierre, ed. *IVe Congrès international de psychologie tenu à Paris en 1900.* Paris: Alcan, 1901.

Jason, Heda, and Dimitri Segal, eds. *Patterns in Oral Literature.* The Hague: Mouton, 1977.

Jefferson, Thomas. *Jefferson's Extracts from the Gospels: "The Philosophy of Jesus" and "The Life and Morals of Jesus."* Princeton: Princeton University Press, 1983.

Jeffery, L. H. *The Local Scripts of Archaic Greece.* Oxford: Clarendon Press, 1961.

———. "ΑΡΧΑΙΑ ΓΡΑΜΜΑΤΑ: Some Ancient Greek Views." In *Europa: Studien zur Geschichte und Epigraphik der frühen Aegaeis*, edited by William Brice, 152–66. Berlin: de Gruyter, 1967.

Johns, Adrian. *The Nature of the Book: Print and Knowledge in the Making*. Chicago: University of Chicago Press, 1998.

Johnson, Edward H. "A Wonderful Invention: Speech Capable of Indefinite Repetition from Automatic Records." *Scientific American* n.s. 37:20 (November 17, 1877), 304.

Jones, Alan H. *Independence and Exegesis: The Study of Early Christianity in the Work of Alfred Loisy (1857–1940), Charles Guignebert (1867–1939), and Maurice Goguel (1880–1955)*. Tübingen: Mohr Siebeck, 1983.

Jousse, Marcel. *Études de psychologie linguistique: Le style oral rythmique et mnémotechnique chez les verbo-moteurs*. (Archives de Philosophie, 2:4.) Paris: Beauchesne, 1925.

———. "Les récitatifs rythmiques de Jésus et de ses apôtres, reconstituées par le R. P. Jousse S.J." Theatre program, Théâtre des Champs-Élysées, 25 April 1929.

———. *Études sur la psychologie du geste. Les rabbis d'Israël: Les récitatifs rythmiques parallèles. I: Genre de la Maxime*. Paris: Spes, 1930.

———. "Les lois psycho-physiologiques du style oral vivant et leur utilisation philologique." *L'Ethnographie*, n.s. 23 (1931): 1–18.

———. "Les outils gestuels de la mémoire dans le milieu ethnique palestinien: le formulisme araméen des récits évangéliques." *L'Ethnographie* n.s. 30 (1935): 1–20.

———. "Le mimisme humain et l'anthropologie du langage." *Revue anthropologique* 57 (1936): 201–215.

———. "Mimisme humain et style manuel." (Travaux du Laboratoire de Rythmo-pédagogie de Paris.) Paris: Geuthner, 1936.

———. "Père, Fils et Paraclet dans le milieu ethnique palestinien." *L'Ethnographie* n.s. 39 (1941): 3–58.

———. "Les formules Targoumiques du 'Pater' dans le milieu ethnique palestinien." *L'Ethnographie* n.s. 42 (1944 [i.e., 1949]): 87–135.

———. "La manducation de la leçon dans le milieu ethnique palestinien." (Travaux du Laboratoire de Rythmo-pédagogie de Paris.) Paris: Geuthner, 1950.

———. "Études de psychologie linguistique. Un génie de la science française: l'abbé J. P. Rousselot et ses découvertes sur le geste oral." Typescript, 18 pp. Collection of the Association Marcel Jousse, Paris.

———. *L'anthropologie du geste*. Paris: Gallimard, 1974.

———. *La manducation de la parole*. Paris: Gallimard, 1975.

———. *Le parlant, la parole et le souffle*. Paris: Gallimard, 1978.

———. *Le style oral*. 2nd edition with notes and commentary by Gabrielle Baron. Paris: Le Centurion, 1981.

———. *Dernières dictées*. Edited by Edgard Sienaert. Paris: Association Marcel Jousse, 1999.

———. *Cours oraux 1931–1957*. CD-ROM. Paris: Association Marcel Jousse, 2004.

———. "Résumés de cours" (unpublished lecture notes). Archives de la Fondation Marcel Jousse, Paris.

Jütte, Robert. *A History of the Senses: From Antiquity to Cyberspace*. Translated by James Lynn. Cambridge: Polity, 2005.

Kahane, Ahuvia. "The First Word of the *Odyssey*." *Transactions of the American Philological Association* 122 (1992): 115–131.

Kahane, Ahuvia, and Martin Mueller, eds. *The Chicago Homer*. Available at http://digital.library.northwestern.edu/homer/.

Kahn, Douglas. "Death in Light of the Phonograph: Raymond Roussel's *Locus Solus*." In *Wireless Imagination: Sound, Radio, and the Avant-Garde*, edited by Douglas Kahn and Gregory Whitehead, 69–103. Cambridge, Mass.: MIT Press, 1992.

Kahn, Gustave. *Premiers poèmes, avec une préface sur le vers libre*. Paris: Mercure de France, 1897.

Karlgren, Bernard, trans. *The Book of Odes: Chinese Text, Transcription and Translation*. Stockholm: Museum of Far Eastern Antiquities, 1950.

Kay, Sarah. "Original Skin: Flaying, Reading, and Thinking in the Legend of Saint Bartholomew and Other Works." *Journal of Medieval and Early Modern Studies* 36 (2006): 35–74.

Kelber, Werner H. *The Oral and the Written Gospel: The Hermeneutics of Speaking and Writing in the Synoptic Tradition, Mark, Paul, and Q*. Philadelphia: Fortress Press, 1983.

———. "Die Fleischwerdung der Wortes in der Körperlichkeit des Textes." In *Materialität der Kommunikation*, edited by Hans Ulrich Gumbrecht and K. Ludwig Pfeiffer, 31–42. Frankfurt am Main: Suhrkamp, 1988.

Kern, Martin. "*Shi jing* Songs as Performance Texts: A Case Study of 'Chu ci' ('Thorny Caltrop)." *Early China* 25 (2000): 49–111.

Kintsch, Walter. *The Representation of Meaning in Memory*. Hillsdale, N.J.: Erlbaum, 1974.

Kiparsky, Paul. "The Role of Linguistics in a Theory of Poetry." *Daedalus* 102 (1973): 231–245.

Kirshenblatt-Gimblett, Barbara. "Folklore's Crisis." *Journal of American Folklore* 111 (1998): 281–327.

Kittler, Friedrich A. *Discourse Networks 1800/1900*. Translated by Michael Metteer and Chris Cullens. Stanford: Stanford University Press, 1990.

———. *Gramophone, Film, Typewriter*. Translated by Geoffrey Winthrop-Young and Michael Wutz. Stanford: Stanford University Press, 1999.

Kittredge, George Lyman. "Introduction." In *English and Scottish Popular Ballads Edited from the Collection of Edward James Child*, edited by Helen

Child Sargent and George Lyman Kittredge, xi–xxxi. Boston: Houghton Mifflin, 1904.

Kohler, K. "Three Trends in Phonetics: The Development of Phonetics as a Discipline in Germany since the Nineteenth Century." In Asher and Henderson, *Towards a History of Phonetics*, 161–178.

Korzybski, Alfred. *Science and Sanity: An Introduction to Non-Aristotelian Systems and General Semantics*. 3rd edition. Lakeville, Conn.: Institute of General Semantics, 1950.

Kress, Gunther. *Before Writing: Rethinking the Paths to Literacy*. New York: Routledge, 1996.

Krois, John Michael, Mats Rosengren, Angela Steidele and Dirk Westerkamp, eds. *Embodiment in Cognition and Culture*. (Advances in Consciousness Research, 71.) Amsterdam: John Benjamins, 2007.

Küster, Ludolphus. *Historia critica Homeri*. Frankfurt am Oder, 1696.

Lafont, Aimé. "Un initiateur en psychologie: Marcel Jousse." *Cahiers du Sud* 95 (1927): 268–299.

Lagrée, Michel, Patrick Harismendy, and Michel Denis, eds. *L'Ouest-Éclair: Naissance et essor d'un grand quotidien régional, 1899–1933*. Rennes: Presses Universitaires de Rennes, 2000.

Lalanne, Léon. *Description et usage de l'abaque, ou Compteur universel qui donne à vue les résultats de tous les calculs d'arithmétique*. Paris: Dubochet, 1845.

La Landelle, Guillaume de. *L'âme et l'ombre d'un navire*. Paris: Chappe, 1861.

Lamberterie, Charles de. "Milman Parry et Antoine Meillet." In Létoublon, *Hommage à Milman Parry*, 9–22.

Landry, Eugène. *La théorie du rythme et le rythme du français déclamé*. Paris: Champion, 1911.

Landry, Lionel. "Le rythme musical." *Revue philosophique* 102 (1926): 223–238.

Langbein, John H. "Historical Foundations of the Law of Evidence: A View from the Ryder Sources." *Columbia Law Review* 96 (1996): 1168–1202.

Laplanche, François. *La crise de l'origine: La science catholique des Évangiles et l'histoire au XXe siècle*. Paris: Albin Michel, 2006.

Laplanche, François, Ilaria Biagioli, and Claude Langlois, eds. *Alfred Loisy cent ans après: Autour d'un petit livre*. Turnhout: Brepols, 2007.

Latour, Bruno, and Steve Woolgar. *Laboratory Life: The Social Construction of Scientific Facts*. Beverly Hills, Calif.: Sage, 1979.

Lauwers, Peter, Marie-Rose Simoni-Aurembou, and Pierre Swiggers, eds. *Géographie linguistique et biologie du langage: Autour de Jules Gilliéron*. Leuven: Peeters, 2002.

Le Blanc, Claudine. "Littératures orales, littérature, et littérature comparée: une discipline pour penser l'oralité littéraire." In *Comparer l'étranger: Enjeux du comparatisme en littérature*, edited by Emilienne Baneth-Nouailhetas and Claire Joubert, 111–127. Rennes: Presses Universitaires de Rennes, 2007.

Lebovics, Herman. *True France: The Wars Over Cultural Identity, 1900–1945.* Ithaca, N.Y.: Cornell University Press, 1992.

Lecoq, Anne-Marie, ed. *La querelle des Anciens et des Modernes, XVIIe–XVIIIe siècles.* Paris: Gallimard, 2010.

Lefèvre, Frédéric. *Marcel Jousse: Une nouvelle psychologie du langage.* (Les Cahiers d'Occident, 10.) Paris: Librairie de France, 1927.

———. "Une nouvelle psychologie du langage." *Le Roseau d'Or: Oeuvres et chroniques* 20 (1927): 3–82.

———. *L'itinéraire philosophique de Maurice Blondel. Propos recueillis par Frédéric Lefèvre.* Paris: Spes, 1928. Reprint, Paris: Aubier-Montaigne, 1966.

———. "La psychologie expérimentale. Une heure avec M. Pierre Janet." *Les nouvelles littéraires,* 17 March 1928.

———. *L'adhésion.* Avignon: Édouard Aubanel, 1943.

Lenoir, Timothy. "Helmholtz and the Materialities of Communication." *Osiris* 9 (1994): 185–207.

———, ed. *Inscribing Science: Scientific Texts and the Materiality of Communication.* Stanford: Stanford University Press, 1998.

Lepore, Jill. "Joe Gould's Teeth." *The New Yorker* (27 July 2015): 48–49.

Leroi-Gourhan, André. *Le geste et la parole, I: Technique et langage. II: La mémoire et les rythmes.* Paris: Albin Michel, 1964–65.

Le Roy Ladurie, Emmanuel. *Montaillou, village occitan de 1294 à 1324.* 2nd edition. Paris: Gallimard, 1985.

Lessing, Gotthold Ephraim. *Philosophical and Theological Writings.* Edited by H. B. Nisbet. Cambridge: Cambridge University Press, 2005.

Létoublon, Françoise. "Le messager fidèle." In Bremer, de Jong, and Kalff, *Homer: Beyond Oral Poetry,* 123–144.

———, ed. *Hommage à Milman Parry. Le style formulaire de l'épopée homérique et la théorie de l'oralité poétique.* Amsterdam: Gieben, 1997.

Lévi-Strauss, Claude. *Anthropologie structurale.* Paris: Plon, 1958.

———. *Le cru et le cuit.* Paris: Plon, 1964.

Lévy-Bruhl, Lucien. *Les fonctions mentales dans les sociétés inférieures.* (1st ed., 1910.) Paris: Presses Universitaires de France, 1951.

Leys, Ruth. *Trauma: A Genealogy.* Chicago: University of Chicago Press, 2000.

Li, Victor. *The Neo-Primitivist Turn: Critical Reflections on Alterity, Culture, and Modernity.* Toronto: University of Toronto Press, 2006.

Liberman, Alvin M., and Ignatius G. Mattingly. "The Motor Theory of Speech Perception Revised." *Cognition* 21 (1985): 1–36.

Liebing, Heinz. "Adolf von Harnack." *Neue deutsche Biographie* 7:688–690.

Littré, Emile. *Dictionnaire de la langue française.* Paris: Hachette, 1873.

Loisy, Alfred Firmin. (As "François Jacobé.") "L'origine du *Magnificat.*" *Revue d'histoire et de littérature religieuses* 2 (1897): 424–432.

———. *L'évangile et l'église.* Paris: Picard, 1902.

———. *Autour d'un petit livre.* Paris: Picard, 1903.

———. *Les évangiles synoptiques.* 2 vols. Ceffonds: Chez l'auteur, 1907–1908.

———. *Simples réflexions sur le décret du Saint-Office 'Lamentabili sane exitu' et sur l'encyclique 'Pascendi dominici gregis.'* 2nd edition. Ceffonds: Chez l'auteur, 1908.

———. *Les livres du Nouveau Testament, traduits du grec en français avec introduction générale et notices.* Paris: Nourry, 1922.

———. "Le style rythmé du Nouveau Testament." *Journal de psychologie* 20 (1923): 405–439.

———. "Review of Jousse, *Le style oral.*" *Journal de psychologie* 22 (1925): 539–542.

———. *La naissance du christianisme.* Paris: Nourry, 1933.

Lomax, Alan. *Mister Jelly Roll: The Fortunes of Jelly Roll Morton, New Orleans Creole and "Inventor of Jazz."* 1950. 2nd edition, Berkeley: University of California Press, 1993.

Lord, Albert Bates. "Homer's Originality: Oral Dictated Texts," *Transactions of the American Philological Association* 84 (1953): 124–34.

———. *The Singer of Tales.* Cambridge, Mass.: Harvard University Press, 1960. 2nd edition, Cambridge, Mass.: Harvard University Press, 2000.

———. *Epic Singers and Oral Tradition.* Ithaca, N.Y.: Cornell University Press, 1991.

———. *The Singer Resumes the Tale.* Edited by Mary Louise Lord. Ithaca, N.Y.: Cornell University Press, 1995.

———, ed. *Serbocroatian Heroic Songs, Collected by Milman Parry,* vol. 1. Cambridge, Mass., and Belgrade: Harvard University Press and Serbian Academy of Sciences, 1954.

Lote, Georges. *Études sur le vers français. Première partie: L'alexandrin d'après la phonétique expérimentale.* 3 vols. Paris: Crès, 1919.

———. *Histoire du vers français.* 3 vols. Paris: Boivin, 1949.

Lowe, N. J. *The Classical Plot and the Invention of Western Narrative.* Cambridge: Cambridge University Press, 2000.

Lü Buwei [attr.]. *Lüshi chunqiu* (The Springs and Autumns of Mr. Lü). In Xu Weiqu, ed., *Lüshi chunqiu jishi.* Beijing: Zhongguo shudian, 1985.

Luriia, Aleksandr Romanovich. *The Mind of a Mnemonist: A Little Book about a Vast Memory.* Translated by Lynn Solotaroff. New York: Basic Books, 1968.

Luzel, François-Marie. "Contes et récits populaires des Bretons Armoricains." *Revue de Bretagne et de Vendée* 13 (1869), 103–108.

Lynn-George, Michael. *Epos: Word, Narrative and the Iliad.* Atlantic Highlands, N.J.: Humanities Press International, 1988.

MacDougall, Robert. "The Structure of Simple Rhythm Forms." *Psychological Review Monograph Supplements* 4 (1903): 309–411.

Maconie, Robin. "The French Connection: Motor Phonetics and Modern Music: Serialism, Gertrude Stein, and Messaien's 'Mode de Valeurs.'" Available at www.jimstonebraker.com/maconie_lingtherory.pdf.

Macpherson, James, trans. [*sic*]. *The Poems of Ossian*. Boston: Crosby, Nichols, Lee & Co., 1861.

Magoun, Francis P., Jr. "The Oral-Formulaic Character of Anglo-Saxon Narrative Poetry." *Speculum* 28 (1953): 446–467.

Mallarmé, Stéphane. *Oeuvres complètes*. Edited by Henri Mondor and G. Jean-Aubry. Paris: Gallimard, 1945.

Mallery, Garrick. *Sign Language Among North American Indians*. (Annual Reports of the Bureau of American Ethnology, vol. 1.) Washington, D.C.: Smithsonian Institution, 1881.

Manovich, Lev. "What Is Digital Cinema?" Available at www.manovich.net/text/digital-cinema.html, 2–3.

"Marcel Jousse: Pour une anthropologie autre." Special number, *Nunc: revue anthropologique* 25 (2011).

"Marcel Jousse: Un' estetica fisiologica." Special number, *Il Cannochiale* 1–3 (2005).

"Marcel Mauss et les techniques du corps." Special number, *Le Portique* 17 (2006).

Marchal, Hugues. "Physiologie et théorie littéraire." (Colloque en ligne: Paul Valéry et l'idée de littérature.) *Fabula: la recherche en littérature*, 8 April 2011. Available at www.fabula.org/colloques/document1416.php. Accessed 24 July 2014.

Marey, Etienne-Jules. *Du mouvement dans les fonctions de la vie, leçons faites au Collège de France*. Paris: Baillière, 1868.

———. *Physiologie expérimentale: Travaux du laboratoire de M. Marey*. 4 vols. Paris: Masson, 1876–80.

———. *La méthode graphique dans les sciences expérimentales et principalement en physiologie et en médecine*. 2nd edition. Paris: Masson, 1885.

———. *Le mouvement*. 1894; reprint, Nîmes: Editions Jacqueline Chambon, 1994.

Marignac, Aloys de. "Esquisse d'une nouvelle méthode de critique homérique." *La Revue du Caire* 5 (1942): 551–561.

Marin, Louis. "Un exemple des rapports entre les études ethniques et anthropologiques: les systèmes de versification." Institut International d'Anthropologie, *IIIe session. Amsterdam, 20–29 septembre 1927*, 493–499.

———. "Le 'phénomène de retour' en anthropologie." *Bulletins et Mémoires de la Société d'Anthropologie de Paris*, series 8, 10 (1939): 73–98.

Maritain, Jacques. *Art et scolastique*. 1920. 2nd edition, Paris: Rouart, 1927.

———. *Antimoderne*. Paris: Éditions de la Revue des Jeunes, 1922.

Marmier, Xavier. "Poésies populaires de nos provinces." *Revue de Paris* 20 (1835): 285–311.

Martin, Henri-Jean. *Histoire et pouvoirs de l'écrit*. Paris: Perrin, 1988.

Martin, Henri-Jean, and Lucien Febvre. *L'Apparition du livre*. Paris: Albin Michel, 1958.

Martin, Richard P. *The Language of Heroes: Speech and Performance in the Iliad.* Ithaca, N.Y.: Cornell University Press, 1989.

Maturana, Humberto, and Francisco Varela. *Autopoiesis and Cognition: The Realization of the Living.* Boston: Reidel, 1980.

Maudsley, Henry. *The Physiology of Mind.* 3rd edition. London: Macmillan, 1876.

Maurras, Charles. *Le chemin de Paradis.* Paris: Calmann-Lévy, 1895.

———. Deleted passages from *Le chemin de Paradis.* Available at http://maurras .net/2012/07/12/preface-du-venin. Accessed 12 July 2014.

Mauss, Marcel. "Fragment d'un plan de sociologie générale descriptive." *Les Annales sociologiques,* serie A (Sociologie générale), fascicule 1 (1934): 1–56.

———. "Les techniques du corps." *Journal de psychologie normale et pathologique* 32 (1936): 271–293. Reprinted in Marcel Mauss, *Sociologie et anthropologie,* 365–386. Paris: Presses Universitaires de France, 1950.

McGann, Jerome. *The Textual Condition.* Princeton: Princeton University Press, 1991.

McLuhan, Marshall. *The Gutenberg Galaxy.* Toronto: University of Toronto Press, 1962.

———. *Understanding Media: The Extensions of Man.* (1964) 2nd edition. Edited by Terrence Gordon. Corte Madera, Calif.: Gingko Press, 2003.

Meier, John. "Einleitung." In *Balladen. Sammlung literarischer Kunst- und Kulturdenkmäler in Entwicklungsreihen,* Reihe 10: *Das deutsche Volkslied,* 1:7–34. Leipzig: Reclam, 1935.

Meillet, Antoine. *Les origines indo-européennes des mètres grecs.* Paris: Presses Universitaires de France, 1923.

———. Review of Jousse, *Le style oral. Bulletin de la Société de Linguistique de Paris* 26 (1925): 5.

———. Review of Jean Paulhan, *Les hain-teny merinas. Bulletin de la Société de Linguistique de Paris* 18 (1913): 356–357.

Meister, Karl. *Die homerische Kunstsprache.* Leipzig: Teubner, 1921.

Menand, Louis. *The Metaphysical Club: A Story of Ideas in America.* New York: Octavo, 2001.

Meschonnic, Henri. *Critique du rythme: Anthropologie historique du langage.* Paris: Verdier, 1982.

———. "Qu'entendez-vous par oralité?" *Langue française* 56 (1982): 6–23.

Meyer-Kalkus, Reinhart. "Work, Rhythm, Dance: Prerequisites for a Kinaesthetics of Media and Arts." 165–181 in John Michael Krois, ed., *Embodiment in Cognition and Culture.* Amsterdam: Benjamins, 2007.

Micale, Mark S., ed. *The Mind of Modernism: Medicine, Psychology, and the Cultural Arts in Europe and America, 1880–1940.* Stanford: Stanford University Press, 2004.

Michon, Pascal. *Rythmes, pouvoir, mondialisation.* Paris: Presses Universitaires de France, 2005.

———. *Marcel Mauss retrouvé: Origines de l'anthropologie du rythme*. Paris: Éditions Rhuthmos, 2010.

———. "Notes éparses sur le rythme comme enjeu artistique, scientifique et philosophique depuis le milieu du XIXe siècle." *Rhuthmos*, 9 November 2012. Available at http://rhuthmos.eu/spip.php?article540. Accessed 3 August 2013.

Miller, Bruce Granville. *Oral History on Trial: Recognizing Aboriginal Narratives in the Courts*. Vancouver: University of British Columbia Press, 2011.

Milne, Anna-Louise. *The Extreme In-Between: Jean Paulhan's Place in the Twentieth Century*. Oxford: Legenda, 2006.

Minchin, Elizabeth. "Scripts and Themes: Cognitive Research and the Homeric Epic." *Classical Antiquity* 11 (1992): 229–241.

———. *Homer and the Resources of Memory: Some Applications of Cognitive Theory to the* Iliad *and the* Odyssey. Oxford: Oxford University Press, 2001.

Momigliano, Arnaldo. *Alien Wisdom: The Limits of Hellenization*. Cambridge: Cambridge University Press, 1975.

Mommsen, Tycho, ed. *Scholia recentiora Thomano-Tricliniana in Pindari Nemea et Isthmia*. Leipzig: Teubner, 1865.

Montandon, George. "Le squelette du Professeur Papillault." *Bulletins et Mémoires de la Société d'anthropologie de Paris* 6 (1935): 5–22.

Montgomery, James. *Lectures on Poetry and General Literature, Delivered at the Royal Institution in 1830 and 1831*. London: Longman, 1833.

Morton, Jelly Roll. *The Complete Library of Congress Recordings*. (CD recording.) Boston: Rounder Records, 2005.

Most, Glenn W. "The Structure and Function of Odysseus' *Apologoi*." *Transactions of the American Philological Association* 119 (1989): 15–35.

Most, Glenn W., Larry Norman, and Sophie Rabau, eds. *Révolutions homériques*. Pisa: Edizioni della Normale, 2009.

Mucchielli, Laurent. *La découverte du social: Naissance de la sociologie en France (1870–1914)*. Paris: La Découverte, 1998.

———. "Sociologie et psychologie en France, l'appel à un territoire commun: Vers une psychologie collective (1890–1940)." *Revue de synthèse* 4.3–4 (1994): 445–483.

Müller-Sievers, Helmut. *The Cylinder: Kinematics of the Nineteenth Century*. Berkeley: University of California Press, 2012.

Murko, Matija. "The Singers and Their Epic Songs." Translated by John Miles Foley. *Oral Tradition* 5 (1990): 107–130.

Murray, Robert W. "Language and Space: The Neogrammarian Tradition." In *Language and Space: An International Handbook of Linguistic Variation*, vol. 1: *Theories and Methods*, edited by Peter Auer and Jürgen Erich Schmidt, 70–79. Berlin: de Gruyter, 2010.

Musa, Aisha Y. *Hadith as Scripture: Discussions on the Authority of Prophetic Traditions in Islam*. New York: Palgrave Macmillan, 2008.

"M.V." "Père Jousse et la maudite conspiration juive." *Manicipium Virginis*, July 2011. Available at http://mancipiumvirginis.blogspot.fr/2011/07/normal-0 -false-false-false-en-us-x-none_2673.html. Accessed 27 August 2012.

"N., L." "L'encyclique *Pascendi dominici gregis.*" In: *Revue néo-scolastique* 14 (1907): 563–567.

Nagler, Michael N. *Spontaneity and Tradition: A Study in the Oral Art of Homer.* Berkeley: University of California Press, 1974.

Nagy, Gregory. *Comparative Studies in Greek and Indic Meter.* Cambridge, Mass.: Harvard University Press, 1974.

———. *The Best of the Achaeans.* Baltimore: Johns Hopkins University Press, 1979.

———. "An Evolutionary Model for the Text Fixation of Homeric Epos." In Foley, *Oral Traditional Literature*, 390–393.

———. *Greek Mythology and Poetics.* Ithaca, N.Y.: Cornell University Press, 1989.

———. *Pindar's Homer: The Lyric Possession of an Epic Past.* Baltimore: Johns Hopkins University Press, 1990.

———. "Homeric Questions." *Transactions of the American Philological Association* 122 (1992): 17–60.

———. *Poetry as Performance: Homer and Beyond.* Cambridge: Cambridge University Press, 1996.

———. *Homer the Preclassic.* Berkeley: University of California Press, 2010.

Nancy, Jean-Luc. *The Inoperative Community.* Translated by Peter Connor, Lisa Garbus, Michael Holland, and Simona Sawhney. Minneapolis: University of Minnesota Press, 1991.

Nemec, Friedrich. "Naumann, Hans." *Neue deutsche Biographie* 18: 769–770.

Neue deutsche Biographie. 25 vols. Berlin: Duncker & Humboldt, 1953–.

Nicolas, Serge. "La mémoire dans l'oeuvre d'Alfred Binet (1857–1911)." *L'Année psychologique* 94 (1994): 257–282.

———. *La mémoire et ses maladies selon Théodule Ribot.* Paris: L'Harmattan, 2002.

Nicolas, Serge, and David J. Murray. "Le fondateur de la psychologie 'scientifique' française: Théodule Ribot (1839–1916)." *Psychologie et histoire* 1 (2000): 1–42.

Nietzsche, Friedrich. *Sämtliche Werke: Kritische Studienausgabe.* Edited by Giorgio Colli and Massimo Montinari. 15 vols. Berlin: DTV / de Gruyter, 1988.

Nirenberg, David. *Anti-Judaism: The Western Tradition.* New York: Norton, 2013.

Norden, Richard. *Die antike Kunstprosa vom VI. Jahrhundert v. Chr. bis in die Zeit der Renaissance.* Leipzig: Teubner, 1915.

Norman, Larry F. *The Shock of the Ancient.* Chicago: University of Chicago Press, 2011.

Ochs, Elinor. "Transcription as Theory." In *Developmental Pragmatics*, edited by Elinor Ochs and Bambi B. Schieffelin, 43–72. New York: Academic Press, 1979.

O'Keeffe, Katherine O'Brien. *Visible Song: Transitional Literacy in Old English Verse*. Cambridge: Cambridge University Press, 1990.

———. "The Performing Body on the Oral-Literate Continuum: Old English Poetry." In Foley, *Teaching Oral Traditions*, 46–58.

———. "Orality and Literacy in Anglo-Saxon England." In *Medieval Oral Literature*, edited by Karl Reichl, 121–140. Berlin: de Gruyter, 2012.

Olrik, Axel. *Principles for Oral Narrative Research*. Translated by Kirsten Wolf and Jody Jensen. Bloomington: Indiana University Press, 1992.

Ombredane, André. *L'aphasie et l'élaboration de la pensée explicite*. Paris: Presses Universitaires de France, 1951.

Ong, Walter J. *The Presence of the Word: Some Prolegomena for Cultural and Religious History*. New Haven: Yale University Press, 1967.

———. *Orality and Literacy: The Technologizing of the Word*. London: Methuen, 1982.

Opie, Iona, and Peter Opie. *The Singing Game*. Oxford: Oxford University Press, 1985.

Opland, Jeff. *Anglo-Saxon Oral Poetry: A Survey of the Traditions*. New Haven: Yale University Press, 1980.

Österreicher, John M. "Pro perfidis Judaeis." *Theological Studies* 8 (1947): 80–96.

Otis, Laura. "The Metaphoric Circuit: Organic and Technological Communication in the Nineteenth Century." *Journal of the History of Ideas* 63 (2002): 105–128.

OuLiPo. *La littérature potentielle*. Paris: Gallimard, 1973.

Pairault, Claude. "Le prophète Marcel Jousse." *Etudes* 359 (September 1983): 231–243.

Pantalony, David. *Altered Sensations: Rudolph Koenig's Acoustical Workshop in Nineteenth-Century Paris*. Dordrecht: Springer, 2009.

Papillault, Georges. *Des instincts à la personnalité morale: Les conditions biopsychologiques de la vie sociale*. Paris: Chahine, 1930.

Paris, Gaston. "Les parlers de France." *Revue des patois gallo-romans* 2 (1888): 161–175.

Parker, Patricia A. "The Metaphorical Plot." In *Metaphor: Problems and Perspectives*, edited by David S. Miall, 135–157. Atlantic Highlands, N.J.: Humanities Press, 1982.

Parry, Adam. "Have We Homer's *Iliad*?" *Yale Classical Studies* 20 (1966): 175–216.

Parry, Milman. *L'épithète traditionnelle dans Homère: Essai sur un problème de style homérique*. Paris: Les Belles Lettres, 1928.

———. *The Making of Homeric Verse*. Edited by Adam Parry. Oxford: Oxford University Press, 1971.

Parry, Milman, and Albert Bates Lord, eds. *Serbocroatian Heroic Songs*, vol. 1: *Novi Pazar: English Translations*. Cambridge, Mass., and Belgrade, Yugoslavia: Harvard University Press and Serbian Academy of Sciences, 1954.

Passanante, Gerard Paul. *The Lucretian Renaissance: Philology and the Afterlife of Tradition*. Chicago: University of Chicago Press, 2011.

Paulhan, Claire, and Bernard Billaud. "Chronologie biographique de Jean Paulhan (1884–1968)." Internet resource, available at www.atelierpdf.com/paulhan.sljp/acrobat/bio/biographiques1–21.pdf. Accessed 5 June 2009.

Paulhan, Jacqueline Frédéric, ed. *Jean Paulhan et Madagascar (1908–1910)*. (*Cahiers Jean Paulhan*, 2.) Paris: Gallimard, 1982.

Paulhan, Jean. *Les hain-teny merinas, poésies populaires malgaches*. Paris: Geuthner, 1913. Reprinted with an introduction by Bernard Baillaud. Paris: Geuthner, 2007.

———. *Les hain-tenys*. Paris: Gallimard, 1939.

———. *Oeuvres complètes de Jean Paulhan*. 5 vols. Paris: Cercle du livre précieux, 1966.

———. "D'un langage sacré." In *Jean Paulhan et Madagascar (1908–1910)*, edited by Jacqueline Paulhan, 312–321. Paris: Gallimard, 1982.

———. *Cahiers Jean Paulhan. 2: Jean Paulhan et Madagascar (1908–1910)*. Paris: Gallimard, 1982.

———. *On Poetry and Politics*. Translated by Jennifer Bajorek, Charlotte Mandell, and Eric Trudel. Chicago: University of Illinois Press, 2008.

———. *Oeuvres complètes, 2: L'art de la contradiction*. Edited by Bernard Baillaud. Paris: Gallimard, 2009.

Perloff, Marjorie. "Screening the Page/Paging the Screen: Digital Poetics and the Differential Text." In *New Media Poetics: Contexts, Technotexts, and Theories*, edited by Adalaide Morris and Thomas Swiss, 143–164. Cambridge, Mass.: MIT Press, 2006.

———. *Unoriginal Genius: Poetry by Other Means in the New Century*. Chicago: University of Chicago Press, 2010.

Pernot, Hubert. "L'abbé Rousselot (1846–1924)." *Revue de phonétique* 5 (1928): 10–23.

Perrault, Charles. *Histoires ou Contes du temps passé*. Edited by Gilbert Rouger. Paris: Garnier, 1967.

———. *Parallèle des Anciens et des Modernes*. 1692–97; reprint, Geneva: Slatkine, 1979.

Perriault, Jacques. *Mémoires de l'ombre et du son: Une archéologie de l'audiovisuel*. Paris: Flammarion, 1981.

Pierrard, Pierre. *Juifs et catholiques français: D'Edouard Drumont à Jacob Kaplan (1886–1994)*. 2nd edition. Paris: Cerf, 1997.

Pisano, Giusy. *Une archéologie du cinéma sonore*. Paris: CNRS Editions, 2004. Available at http://books.openedition.org/editionscnrs/2715. Accessed 31 July 2013.

Pop, Andrei. *Antiquity, Theatre, and the Painting of Henry Fuseli.* New York: Oxford University Press, 2015.

Porter, James. "'Bring Me the Head of James MacPherson': The Execution of Ossian and the Wellsprings of Folkloristic Discourse." *Journal of American Folklore* 114 (2001): 396–425.

Poulat, Emile. Review of Gabrielle Baron, *Mémoire vivante. Archives des sciences sociales des religions* 53 (1982): 238–239.

———. *Histoire, dogme et critique dans la crise moderniste.* 3rd edition. Paris: Albin Michel, 1996.

Pound, Ezra. *Polite Essays.* London: Faber & Faber, 1937.

Powell, Barry B. *Homer and the Origin of the Greek Alphabet.* Cambridge: Cambridge University Press, 1991.

Propp, Vladimir. *Theory and History of Folklore.* Edited by Anatoly Liberman; translated by Ariadna Y. Martin and Richard P. Martin. Minneapolis: University of Minnesota Press, 1984.

Puech, Christian. "Langage intérieur et ontologie linguistique à la fin du XIXe siècle." *Langue française* 132 (2001): 26–47.

Quintilianus, M. Fabius. *Institutionis oratoriae libri duodecim.* Edited by Edward Bonnell. 2 vols. Leipzig: Teubner, 1861.

Rabau, Sophie. "Pour une poétique de l'interpolation." *Fabula-LhT* 5, "Poétique de la philologie," November 2008, available at www.fabula.org/lht/5/rabau .html. Accessed 22 July 2014.

Rabinbach, Anson. *The Human Motor: Energy, Fatigue, and the Origins of Modernity.* Berkeley: University of California Press, 1992.

Radloff, Wilhelm. "Samples of Folk Literature from the North Turkic Tribes." Translated by Gudrun Böttcher Sherman and Adam Brooke Davis. *Oral Tradition* 5 (1990): 73–90.

Reichl, Karl. *Singing the Past: Turkic and Medieval Heroic Poetry.* Ithaca, N.Y.: Cornell University Press, 2000.

———, ed. *Medieval Oral Literature.* Berlin: de Gruyter, 2012.

Renan, Ernest. *L'Avenir de la science: pensées de 1848.* Paris: Calmann-Lévy, 1890. Translated by Albert D. Vandam and C. B. Pitman as *The Future of Science: Ideas of 1848.* London: Chapman and Hall, 1891.

Renouard, Antoine Augustin. *Annales de l'imprimerie des Estienne, ou histoire de la famille des Estienne et de ses éditions.* Paris: Renouard, 1837.

Rey, André-Louis, ed. *Patricius, Eudocie, Optimus, Côme de Jérusalem: Centons homériques (Homerocentra).* (Sources chrétiennes, 437.) Paris: Les Éditions du Cerf, 1998.

"Rhythmocatechist." *Time,* 6 November 1939, 54.

Ribot, Théodule. "Les mouvements et leur importance psychologique." *Revue philosophique* 8 (1879): 371–386.

———. *Les maladies de la mémoire.* Paris: Baillière, 1881. Reprint, Paris: Alcan, 1921.

———. *Les maladies de la volonté*. Paris: Baillière, 1883. Reprint, Paris: Alcan, 1921.

———. *Les maladies de la personnalité*. Paris: Alcan, 1885. Reprint, Paris: Alcan, 1921.

———. *Psychologie de l'attention*. Paris: Alcan, 1889.

———. *La vie inconsciente et les mouvements*. Paris: Alcan, 1914.

Richet, Charles. "De l'influence des mouvements sur les idées." *Revue philosophique* 8 (1879): 610–615.

Robiglio, Arianna. *In principio era il gesto: Introduzione alla pedagogia di Marcel Jousse*. Pisa: Servizio editoriale universitario, 2000.

Robins, R. H. *A Short History of Linguistics*. Bloomington: Indiana University Press, 1968.

Rodrigues, Jacob Hippolyte. *Les origines du sermon sur la montagne*. Paris: Lévy, 1868.

Rosapelly, Charles. "Inscription des mouvements phonétiques." In *Physiologie expérimentale: Travaux du laboratoire de M. Marey*, 2:109–131. Paris: Masson, 1876.

———. "Nouvelles recherches sur le rôle du larynx dans les consonnes sourdes et sonores." *Mémoires de la Société de linguistique de Paris* 9 (1896): 488–499.

———. "Valeur relative de l'implosion et de l'explosion dans les consonnes occlusives." *Mémoires de la Société de linguistique de Paris* 10 (1898): 347–363.

———. "Analyse graphique de la consonne: Sa division en trois temps." *Mémoires de la Société de linguistique de Paris* 10 (1898): 71–79.

Rosen, Jody. "Researchers Play Tune Recorded Before Edison." *New York Times*, 27 March 2008.

Rousseau, Jean-Jacques. *Essai sur l'origine des langues, où il est parlé de la mélodie, et de l'imitation musicale*. In *Oeuvres complètes*, 5:371–429. Edited by Bernard Gagnebin and Marcel Raymond. Paris: Gallimard, 1995.

Rousseau, Pascal. "Figures de déplacement: L'écriture du corps en mouvement." *Exposé* 2 (1995): 86–97.

Roussel, Raymond. *Locus Solus*. Paris: Gallimard, 1963.

Rousselot, Pierre Jean. *Les modifications phonétiques du langage, étudiées dans le patois d'une famille de Cellefrouin (Charente)*. Paris: Welter, 1891; also published in *Revue des patois gallo-romans* 4 (1891): 65–208, 5 (1892): 209–380, 5 supp. (1893): 9–48.

———. "La méthode graphique appliquée à la recherche des transformations inconscientes du langage." *Revue des patois gallo-romans* 4 (1891): 209–213.

———. "Classification des voyelles orales: désignation des nuances de timbre et signes pour les représenter." *Revue de phonétique* 1 (1911): 17–32.

———. *La phonétique expérimentale: Leçon d'ouverture du cours professé au Collège de France*. Paris: Boivin, 1922.

———. *Principes de phonétique expérimentale*. 2 vols. Paris: Welter, 1897. 2nd edition, Paris: Didier, 1924.

Rubin, David C. *Memory in Oral Traditions: The Cognitive Psychology of Epic, Ballads, and Counting-Out Rhymes*. New York: Oxford University Press, 1995.

Saussure, Ferdinand de. "Notes inédites de F. de Saussure." *Cahiers Ferdinand de Saussure* 12 (1954): 49–71.

———. *Cours de linguistique générale*. Edited by Charles Bally, Albert Sechehaye, and Albert Riedlinger. Paris: Payot, 1972 (1916).

———. *Ecrits de linguistique générale*. Edited by Simon Bouquet and Rudolf Engler. Paris: Gallimard, 2002.

———. *Course in General Linguistics*. Edited by Perry Meisel and Haun Saussy. Translated by Wade Baskin. New York: Columbia University Press, 2011.

Saussy, Haun. "Writing in the *Odyssey*: Eurykleia, Parry, Jousse and the Opening of a Letter from Homer." *Arethusa* 29 (1996): 299–338.

———. "Rhyme, Repetition and Exchange in the *Book of Odes*." *Harvard Journal of Asiatic Studies* 57 (1997): 519–542.

———. "The Refugee Speaks of Parvenus and their Beautiful Illusions: A Rediscovered 1934 Text by Hannah Arendt." *Critical Inquiry* 40 (2013): 1–14.

Scheffer, Pierre, SJ. "Marcel Jousse (1886–1961) ou le service de la Parole, humaine et divine." *Études théologiques et religieuses* 63 (1988): 367–378.

Schlanger, Nathan. "Le fait technique total. La raison poétique et les raisons de la poétique dans l'oeuvre de Marcel Mauss." *Terrain* 16 (1991), 114–130. Available at http://terrain.revues.org/3003. Accessed 9 August 2012.

Schleiermacher, Friedrich Daniel Ernst. *Über die Religion: Reden an die Gebildeten unter ihren Verächtern* (1799). In *Schleiermachers Werke*, 4:207–399. Edited by Otto Braun and Johannes Bauer. Leipzig: Meiner, 1911.

Schloesser, Stephen. *Jazz Age Catholicism: Mystic Modernism in Postwar Paris, 1919–1933*. Toronto: University of Toronto Press, 2005.

Schmandt-Besserat, Denise. *Before Writing*. 2 vols. Austin: University of Texas Press, 1992.

Schmidt, Carl Eduard. *Parallel-Homer, oder Index aller homerischen Iterati in lexicalischer Ordnung*. Göttingen: Vandenhoeck & Ruprecht, 1885.

Schmidt, Karl-Ludwig. *Der Rahmen der Geschichte Jesu: literarkritische Untersuchungen zur ältesten Jesusüberlieferung* (Berlin: Trowitzsch, 1919).

Schwob, Marcel. *Oeuvres*. Edited by Alexandre Gefen. Paris: Les Belles Lettres, 2002.

Scott, Clive. "Re-Conceiving Voice in Modern Verse." *Comparative Critical Studies* 5 (2008): 5–20.

Scott de Martinville, Edouard-Léon. *The Phonautographic Manuscripts of Edouard-Léon Scott de Martinville*. Edited and translated by Patrick Feaster. 2009, 2011. Available at www.phonozoic.net/fs/Phonautographic-Manuscripts.pdf. Accessed 28 July 2013.

———. "Brevet d'Invention (1857) and Certificat d'Addition (1859)." Available at www.firstsounds.org/publications/facsimiles/FirstSounds_Facsimile_02.pdf. Accessed 28 July 2013.

Sealey, Raphael. "From Phemios to Ion." *Revue des études grecques* 70 (1957): 312–355.

Sebban, Joël. "La genèse de la 'morale judéo-chrétienne': Etude sur l'origine d'une expression dans le monde intellectuel français." *Revue de l'histoire des religions* 229 (2012): 85–118.

Sebeok, Thomas, ed. *Style in Language.* Cambridge, Mass: MIT Press, 1960.

Sébillot, Paul. *Littérature orale de la Haute-Bretagne.* Paris: Maisonneuve, 1881.

Sériot, Patrick. *Structure et totalité: Les origines intellectuelles du structuralisme en Europe centrale et orientale.* Paris: Presses Universitaires de France, 1999. 2nd edition, Paris: Lambert-Lucas, 2012.

Serry, Hervé. *Naissance de l'intellectuel catholique.* Paris: La Découverte, 2004.

Severyns, Albert. Review of Milman Parry, *L'épithète traditionnelle dans Homère. Revue belge de philologie et d'histoire* 8 (1929): 881–885.

Sharp, Cecil J. *English Folk-Song: Some Conclusions.* London: Simpkin, 1907.

Shive, David. *Naming Achilles.* Oxford: Oxford University Press, 1987.

Sieburth, Richard. "The Work of Voice in the Age of Mechanical Reproduction." 2007. Available at nn.edu/pennsound/x/text/Sieburth-Richard_Pound.html. Accessed 1 July 2013.

Sienaert, Edgard Richard. "Marcel Jousse: The Oral Style and the Anthropology of Gesture," *Oral Tradition* 5 (1990): 91–106.

Silverstein, Michael. "From Baffin Island to Boasian Induction: How Anthropology and Linguistics Got Into Their Interlinear Groove." In Regna Darnell, ed., *The Franz Boas Papers*, vol. 1. *Franz Boas as Public Intellectual: Theory, Ethnography, Activism*, 83–127. Lincoln: University of Nebraska Press, 2015.

[Sima Qian] Ssu-ma Ch'ien, *The Grand Scribe's Records, Vol. 1: The Basic Annals of Pre-Han China.* Edited by William H. Nienhauser Jr.; translated by Tsai-fa Cheng, Zongli Lu, William H. Nienhauser Jr., and Robert Reynolds. Bloomington: Indiana University Press, 1994.

Simondon, Gilbert. *Du mode d'existence des objets techniques.* Paris: Aubier, 1958.

Simonsuuri, Kirsti. *Homer's Original Genius: Eighteenth-Century Notions of the Early Greek Epic, 1688–1798.* Cambridge: Cambridge University Press, 1979.

Skafte Jensen, Minna. *The Homeric Question and the Oral-Formulaic Theory* (Opuscula Graecolatina, 20). Copenhagen: Museum Tusculanum, 1980.

Smith, Richard Cándida. *Mallarmé's Children: Symbolism and the Renewal of Experience.* Berkeley: University of California Press, 1999.

Snyder, Joel. "Visualization and Visibility." In *Picturing Science, Producing Art*, edited by Carolina Jones, Peter Galison, and Amy Slater, 379–397. New York: Routledge, 1998.

Soriano, Marc. *Les Contes de Perrault: Culture savante et traditions populaires.* Paris: Gallimard, 1968.

Spencer, Herbert. *The Philosophy of Style.* New York: Appleton, 1882 (1852).

Spire, André. *Plaisir poétique et plaisir musculaire: essai sur l'évolution des techniques poétiques.* Paris and New York: José Corti / S. F. Vanni, 1949. 2nd edition, Paris: José Corti, 1986.

Stanley, Keith. *The Shield of Homer.* Princeton: Princeton University Press, 1993.

Steiner, Peter, ed. *The Prague School: Selected Writings, 1929–1946.* Austin: University of Texas Press, 1982.

Stephanus, Henricus (Henri Estienne II). *Homeri et Hesiodi certamen.* Geneva, 1573.

Sterne, Jonathan. *The Audible Past: Cultural Origins of Sound Reproduction.* Durham, N.C.: Duke University Press, 2003.

———. "The Theology of Sound: A Critique of Orality." *Canadian Journal of Communication* 36 (2011): 207–225.

Stetson, Raymond Herbert. "Rhythm and Rhyme." *Psychological Review Monograph Supplements* 4 (1903): 413–466.

———. "A Motor Theory of Rhythm and Discrete Succession." *Psychological Review* 12 (1905): 250–270.

———. *Motor Phonetics: A Study of Speech Movements in Action.* (Archives néerlandaises de phonétique expérimentale, 3.) The Hague: Nijhoff, 1928.

———. *Bases of Phonology.* Oberlin, Ohio: Oberlin College, 1945.

Stewart, Jude. *Patternalia: An Unconventional History of Polka Dots, Stripes, Plaid, Camouflage and Other Graphic Patterns.* New York: Bloomsbury, 2015.

Stiegler, Bernard. *La technique et le temps, I: La Faute d'Épiméthée.* Paris: Galilée, 1994.

Stock, Brian. *The Implications of Literacy: Written Language and Models of Interpretation in the Eleventh and Twelfth Centuries.* Princeton: Princeton University Press, 1983.

———. *Listening for the Text: On the Uses of the Past.* Philadelphia: University of Pennsylvania Press, 1996.

Stolz, Benjamin A., and Richard S. Shannon, eds. *Oral Literature and the Formula.* Ann Arbor: Center for the Coordination of Ancient and Modern Studies, 1976.

Stössinger, Felix. "Feuilleton: Vom französischen Judentum." *Die Wahrheit, Jüdische Wochenschrift* 50.44 (2 November 1934): 2.

Strauss, David Friedrich. *Das Leben Jesu für das deutsche Volk.* Leipzig: Brockhaus, 1864.

Sutton, Michael. *Nationalism, Positivism and Catholicism: The Politics of Charles Maurras and French Catholics, 1890–1914.* Cambridge: Cambridge University Press, 1982.

Svenbro, Jesper. *Phrasikleia: Anthropologie de la lecture en Grèce ancienne.* Paris: La Découverte, 1988.

Swindle, P. F. "On the Inheritance of Rhythm." *American Journal of Psychology* 24 (1913): 180–203.

Syrotinski, Michael. *Defying Gravity: Jean Paulhan's Interventions in Twentieth-Century French Intellectual History.* Albany: State University of New York Press, 1998.

Szendy, Peter. *Écoute: une histoire de nos oreilles.* Paris: Minuit, 2001.

Taine, Hippolyte. *De l'intelligence.* 2 vols. 11th edition, Paris: Hachette, 1906.

Tamine, Joëlle Gardes. "Le vers de *La Légende des Siècles.*" In *Victor Hugo, La Légende des Siècles, première série,* edited by André Guyaux and Bertrand Marchal, 101–120. Paris: Presses Universitaires de la Sorbonne, 2002.

Tate, Aaron Phillip. "Matija Murko, Wilhelm Radloff, and Oral Epic Studies." *Oral Tradition* 26 (2011): 329–352.

Taupin, René. *L'influence du Symbolisme français sur la poésie américaine (de 1910 à 1920).* Paris: Champion, 1929.

Taylor, Charles. *Philosophical Arguments.* Cambridge, Mass.: Harvard University Press, 1995.

Tchougounnikov, Sergei, and Céline Trautmann-Waller, eds. *Pëtr Bogatyrëv et les débuts du Cercle de Prague.* Paris: Presses Sorbonne Nouvelle, 2012.

Tel Quel. *Théorie d'ensemble.* Paris: Seuil, 1968.

Temple, Kathryn. "Johnson and Macpherson: Cultural Authority and the Construction of Literary Property." *Yale Journal of Law and the Humanities* 5 (1993): 355–387.

Tertullian. *De praescriptione haereticorum.* Edited and translated by R. F. Refoulé as *Traité de la prescription contre les hérétiques.* Paris: Editions du Cerf, 1957.

Testenoire, Pierre-Yves. "Marcel Jousse, Antoine Meillet et l'oralité poétique." Manuscript, September 2011.

———. "Littérature orale et sémiologie saussurienne." In *En quoi Saussure peut-il nous aider à penser la littérature?* Edited by Sandrine Bédouret-Larraburu and Gisèle Prignitz, 61–77. (Linguistique et littérature, 1.) Pau: Presses Universitaires de Pau, 2012.

Teston, Bernard. "L'oeuvre d'Etienne-Jules Marey et sa contribution à l'émergence de la phonétique dans les sciences du langage." *Travaux Interdisciplinaires du Laboratoire Parole et Langage* 23 (2004): 237–266.

Tharaud, Jérôme, and Jean Tharaud. "L'an prochain à Jérusalem, 1: Les trois prières." *Revue des deux mondes* 94 (1924): 756–785.

Thomas, Adolphe V. "L'anthropologie du geste et les proverbes de la Terre." *Revue anthropologique* 51 (1941): 164–194.

Tiffany, Daniel. *Radio Corpse: Imagism and the Cryptaesthetic of Ezra Pound.* Cambridge, Mass.: Harvard University Press, 1995.

Tihanov, Galin. "When Eurasianism Met Formalism: An Episode from the History of Russian Intellectual Life in the 1920s." *Die Welt der Slaven* 48 (2003): 359–382.

Tincq, Henri. *Dieu en France: Mort et résurrection du catholicisme.* Paris: Calmann-Lévy, 2003.

Torgovnick, Marianna. *Gone Primitive: Savage Intellects, Modern Lives.* Chicago: University of Chicago Press, 1991.

Tourtoulon, Charles de, and Octave Bringuier. *Étude sur la limite géographique de la langue d'oc et de la langue d'oïl, avec une carte.* Paris: Imprimerie Nationale, 1876.

Troubetzkoy, N. S. "La phonologie actuelle." *Journal de psychologie normale et pathologique* 30 (1933): 227–246.

———. *Principes de phonologie.* Trans. J. Cantineau. Paris: Klincksieck, 1957.

———. "Bericht von Prof. Dr. N. Troubetzkoy." *Actes du deuxième Congrès International de Linguistes,* 120–129.

———. *Studies in General Linguistics and Language Structure.* Edited by Anatoly Liberman, tr. Marvin Taylor and Anatoly Liberman. Durham, N.C.: Duke University Press, 2001.

Trumpener, Katie. *Bardic Nationalism: The Romantic Novel and the British Empire.* Princeton: Princeton University Press, 1997.

Tufte, Edward R. *The Visual Display of Quantitative Information.* Cheshire, Conn.: Graphics Press, 1983.

Usher, Mark D. *Homeric Stitchings: The Homeric Centos of the Empress Eudocia.* Lanham, Md.: Rowman and Littlefield, 1998.

———, ed. *Homerocentones Eudociae Augustae.* Stuttgart: Teubner, 1999.

Vaïsse, Léon. "Notes pour server à l'histoire des machines parlantes." *Mémoires de la Société de linguistique de Paris* 3 (1878): 257–268.

Valéry, Paul. *Oeuvres.* 2 vols. Edited by Jean Hytier. Paris: Gallimard, 1957–60.

van Biervliet, J. J. "Images sensitives et images motrices." *Revue philosophique de la France et de l'étranger* 44 (1897): 113–128.

van Gennep, Arnold. *La question d'Homère: Les poèmes homériques, l'archéologie et la poésie populaire.* Paris: Mercure de France, 1909.

Varela, Francisco J, Evan Thompson, and Eleanor Rosch. *The Embodied Mind: Cognitive Science and Human Experience.* Cambridge, Mass.: MIT Press, 1993.

Vaugeois, Henri. "L'Action française." *L'Action française* 1 (1899): 7–25.

Vermes, Geza. *The Religion of Jesus the Jew.* Minneapolis: Fortress Press, 1993.

Vico, Giambattista. *La Scienza nuova.* Edited by Paolo Rossi. Milan: Rizzoli, 1977.

Virolleaud, Charles. "Notice sur la vie et les travaux de M. Aimé Puech." *Comptes-rendus des séances de l'Académie des inscriptions et belles-lettres* 91 (1947): 136–151.

Vitry, Alexandre de. "Catholicisme et représentation: de Pie IX à la littérature exotopique." 2009. Available at http://etudes-romantiques.ish-lyon.cnrs.fr/wa_files/AlexandredeVitry.pdf. Accessed 4 July 2013.

von Kleist, Heinrich. "Über das Verfertigen der Gedanken beim Reden." In *Werke in einem Band,* 810–814. Munich: Carl Hanser, 1966.

Vulliaud, Paul. "Le style des Evangiles et les théories du Père Jousse." *Mercure de France* 787 (1 April 1931): 77–99.

Vygotsky, Lev Semyonovich. *Mind in Society: The Development of Higher Psychological Processes.* Edited by Michael Cole, Vera John-Steiner, Sylvia Scribner, and Ellen Souberman. Cambridge, Mass.: Harvard University Press, 1978.

Wace, A. J. B., and F. H. Stubbings, eds. *A Companion to Homer*. New York: Macmillan, 1962.

Wallaschek, R. "On the Difference of Time and Rhythm in Music." *Mind* n.s. 4 (1895): 28–35.

Wang, C. H. *The Bell and the Drum: Shih Ching as Formulaic Poetry in an Oral Tradition*. Berkeley: University of California Press, 1974.

Warburg, Aby. *L'atlas Mnémosyne*. Translated by Sacha Zilberfarb. Paris: L'Écarquillé, 2012.

Weber, Eugen. *Action Française: Royalism and Reaction in Twentieth-Century France*. Stanford: Stanford University Press, 1962.

White, Jeffrey A. "Bellerophon in the 'Land of Nod': Some Notes on *Iliad* 6.153–211." *American Journal of Philology* 103 (1982): 119–127.

Whitman, Cedric. *Homer and the Heroic Tradition*. Cambridge, Mass.: Harvard University Press, 1958.

Wilamowitz-Moellendorff, Ulrich von. *Homerische Untersuchungen* (Philologische Untersuchungen, 7). Berlin: Weidmann, 1884.

———. *Die Heimkehr des Odysseus: Neue Homerische Untersuchungen*. Berlin: Weidmann, 1927.

Wolf, Friedrich August. *Prolegomena ad Homerum*. Edited by Rudolf Peppmüller. Hildesheim: Olms, 1963.

———. *Prolegomena to Homer*. Translated by Anthony Grafton, Glenn W. Most, and James E. G. Zetzel. Princeton, N.J.: Princeton University Press, 1985.

Wood, Robert. *An Essay on the Original Genius and Writings of Homer*. London, 1775; 2nd edition 1767.

Woodworth, R. S. "Non-Sensory Components of Sense Perception." *Journal of Philosophy, Psychology and Scientific Methods* 4 (1907): 169–176.

Woolf, D. R. "The 'Common Voice': History, Folklore, and Oral Tradition in Early Modern England." *Past and Present* 120 (1988): 26–52.

Wundt, Wilhelm. *Völkerpsychologie, eine Untersuchung der Entwicklungsgesetze von Sprache, Mythos und Sitte*. 3 vols. Leipzig: Engelmann, 1900.

———. *The Language of Gestures*. The Hague: Mouton, 1973.

Young, Stephen E. *Jesus Tradition in the Apostolic Fathers: Their Explicit Appeals to the Words of Jesus in Light of Orality Studies*. Tübingen: Mohr Siebeck, 2011.

Young, Thomas Paul. Review of Jousse, *Le style oral*. *Journal of Philosophy* 23 (1926): 276–277.

Zumthor, Paul. *Essai de poétique médiévale*. Paris: Seuil, 1972.

———. "Le rythme dans la poésie orale." *Langue française* 56 (1982): 114–127.

———. *Introduction à la poésie orale*. Paris: Seuil, 1983.

———. *La lettre et la voix: De la "littérature" médiévale*. Paris: Seuil, 1987.

Zuo Yan. "Classical Chinese Verse-Grammar: Coexisting Sub-grammars and Formal Grounding." PhD dissertation, University of Tilburg, 2002.

Index

VERBAL ARTS : STUDIES IN POETICS

Lazar Fleishman and Haun Saussy, series editors

Kiene Brillenburg Wurth, *Between Page and Screen: Remaking Literature Through Cinema and Cyberspace*

Jacob Edmond, *A Common Strangeness: Contemporary Poetry, Cross-Cultural Encounter, Comparative Literature*

Christophe Wall-Romana, *Cinepoetry: Imaginary Cinemas in French Poetry*

Marc Shell, *Talking the Walk & Walking the Talk: A Rhetoric of Rhythm*

Ryan Netzley, *Lyric Apocalypse: Milton, Marvell, and the Nature of Events*

Ilya Kliger and Boris Maslov (eds.), *Persistent Forms: Explorations in Historical Poetics*. Foreword by Eric Hayot

Ross Chambers, *An Atmospherics of the City: Baudelaire and the Poetics of Noise*

Haun Saussy, *The Ethnography of Rhythm: Orality and Its Technologies*. Foreword by Olga Solovieva